ALL ABOUT HOME SATELLITE TELEVISION

ALL ABOUT HOME SATELLITE TELEVISION

BY RICK COOK & FRANK VAUGHAN

TAB BOOKS Inc.
BLUE RIDGE SUMMIT, PA. 17214

FIRST EDITION

THIRD PRINTING

Printed in the United States of America

Library of Congress Cataloging in Publication Data

Cook, Rick.
All about home satellite television.

Includes index.
1. Earth stations (Satellite telecommunication)—
Amateurs' manuals. 2. Home video systems.
I. Vaughn, Frank. II. Title.
TK9960.C66 1983 621.388 82-19304
ISBN 0-8306-0519-3
ISBN 0-8306-1519-9 (pbk.)

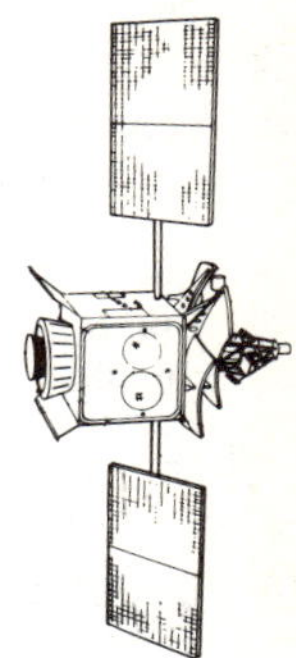

Contents

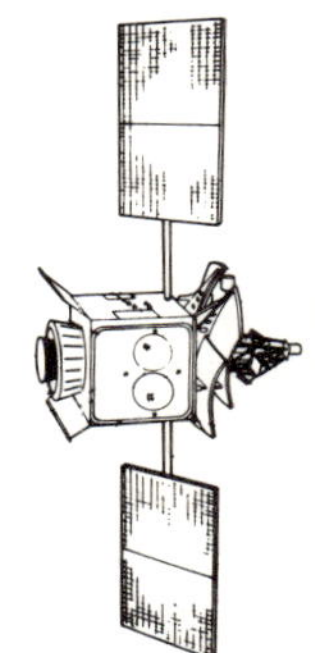

Introduction

It has been about five years since the first experimenter cobbled a mixture of surplus and commercial parts and started picking up television signals from a communications satellite. It has been less time than that since Nieman-Marcus featured a complete home earth station in its Christmas catalog for $25,000.

That was more of a joke than a serious business offer. Every year Nieman-Marcus includes one or two outrageous items in its catalog for publicity purposes. The joke offering of a few years ago has become the basis of a rapidly growing industry.

No one knows how many people have backyard satellite receivers today. The estimates range from 12,000 to 100,000, with a consensus centering on 30,000 to 40,000. We do know that the numbers are increasing rapidly. By next year that figure may well have doubled.

The owners of home earth stations paid a lot less than $25,000 for them. Today you can install your own earth station for less than $5,000 nearly anywhere in the United States. In many places you can do it for between $2,000 and $3,000. Those prices are falling rapidly, and the capability of the equipment is expanding, too. In a few years it may be possible to put in a complete system for $1,000 or less.

In many ways home satellite television is where home computers were about five years ago. Like home computers, home satellite television has the potential to grow explosively. It also has the potential to change our lives by redefining the nature of the television audience and letting television fulfill needs it never could before.

The short-term promise of home satellite television is more kinds of programming than you could ever get without it. If first-run movies, sporting events, and religious and children's programming are broadcast on television, it is available from a satellite.

The longer-term promise is that home satellite television will reshape television. It has the potential to convert television into a truly national medium and to provide programming for specialized audiences in a way that is simply not possible with conventional broadcast television. It's not that home satellite television improves the quality of television; it just gives the viewer so much more choice that the "inferior" programming (however you define that) can be ignored. No matter what your interest, it is there in abundance.

How did we end up with television via satellite anyway? What can it do? What does it take to set up a home earth station? Is it worth it for you?

That is what this book is about. You will have to answer the last question for yourself, but the others we will try to answer for you. In the course of answering them, we are going to cover a lot of territory—from the depths of space to electronics laboratories in Japan to a village in India and the halls of the United States Congress.

Because satellite television is a communications medium and, like all communications media, has social and political dimensions, we are going to spend some time discussing social and political issues. Home satellite television is so new that many of the key decisions have not yet been made—or have only been made very recently. That makes the political dimension particularly important.

Because the medium is so new and is changing so rapidly, it is hard for any book to keep up with it. Much of the information in this book is subject to change with new technical advances, new court cases, and changes in the law. We have done our best to make everything accurate but it is entirely possible that some things will be changed by the time this book is published.

Chapter 1

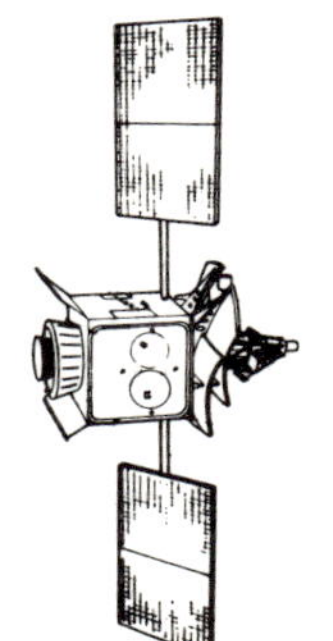

The History of Satellite Television

A science fiction writer first thought seriously about the possibility of communicating via an artificial satellite. The writer was Arthur C. Clarke, and the time was 1945. As a space enthusiast and a member of the British Interplanetary Society, Clarke was already thinking about ways to put objects into space and what they could do once they got there. Clarke was also familiar with radar and the possibilities of microwaves from his wartime work.

Even in the 1940s the idea of interplanetary communication by radio or television was common in science fiction. For example, an American writer named George O. Smith had written a series of stories about an interplanetary radio network that used an artificial satellite of Venus as a relay point. The writers that had really looked at the idea believed such transmissions would be extremely difficult. All the radio frequencies they were familiar with were strongly reflected by the *Heaviside layer* of ionized gases at the upper reaches of the earth's atmosphere. It seemed certain that only a very powerful signal could reach space.

Clarke was the first to realize that the development of microwave radar during World War II changed everything. *Microwaves*, radio waves with a frequency of 1000 megahertz and higher, passed right through the Heaviside layer. As a result of wartime work on radar, microwave frequencies were now accessible. The techniques and equipment necessary to work with them had been developed, and what had been nearly imposssible in 1939 was being done as a routine matter of technical practice by 1945.

Not long after the war ended, the United States Army demonstrated the capabilities of the new technology by bouncing a radar signal off the moon. Clarke proposed orbiting a satellite to send and receive from earth. Such a device would be more reliable and less subject to interference than conventional shortwave transmissions, he said, and it could reach areas that weren't served by undersea cables.

Ironically, when Clarke attempted to patent the idea, he was turned down on the grounds that he couldn't reduce it to practice. Because he couldn't build a working communications satellite, he was denied a patent. Clarke returned to writing science fiction where he has continued to excel. Arthur C. Clarke lives today in Sri Lanka (formerly Ceylon) and uses a private satellite earth station to receive television programs in a country that otherwise lacks television.

THE FIRST SATELLITES

There is a considerable distance between the concept of a communications satellite and a system that can actually relay messages. It took nearly 20 years to go from Clarke's description in an English magazine to the first satellite in commercial service, and another decade to go from that to home reception of satellite television broadcasts. One of the next steps on the road to satellite television was taken by British astronomers who were interested in receiving messages from space—years before there were any satellites in space to send them.

The astronomers were radio astronomers. The messages they wanted to receive came from the stars themselves, not from man-made objects. Wartime advances in radio made radio astronomy practical, and all over the world astronomers were setting up radio telescopes.

Among them was Bernard Lovell, who established the Nuffield Radio Astronomy Laboratories in conjunction with the University of Manchester at Macclesfield, Chesire, England. Most people know the facility by the name of its largest telescope—Jodrell Bank.

Like most radio astronomers, Lovell and his colleagues started with simple antennas like the ones used to receive radio broadcasts. These antennas brought in signals and a lot of electronic noise. Resolution was poor, and it was hard to tell just where in the sky a signal was coming from with such a simple antenna.

Work was begun in 1952 on a steerable dish radio telescope. Although no one thought of it that way, Lovell and his colleagues were building the first Earth station.

The Jodrell Bank may have been an earth station, but it was not something you could set up in your backyard. The dish was 250 feet in diameter and turned on turret mechanisms out of an old battleship. The legs that supported the dish were mounted on railway flatcars running on a circular track. The design let the big dish cover virtually the entire visible sky. The size of the anntenna and the associated specially designed electronics gave Jodrell Bank a sensitivity and resolution far beyond any previous instrument.

It was the second mission that made Jodrell Bank famous. The telescope was finished in 1957, just before the Russians launched Sputnik I. Mankind had entered the space age. The giant Jodrell Bank station was one of the few in existence that could track this planet's first artificial satellite.

The signals coming down from Sputnik I after its launch on October 4, 1957, were on a known frequency and about as complicated as the chirping of a cricket. Like a cricket's chirp, Sputnik's "beep-beep" told the listener its location, and it speeded up or slowed down according to the temperature. The satellite's transmitter didn't have much power. While anyone with a fairly good ham radio receiver tuned to the right frequency could hear Sputnik. Better equipment was needed to track it accurately. Soon there was a worldwide network of steerable dish earth stations to track satellites.

THE EARLY COMSATS

The first satellites to go into orbit were combination propaganda showpieces and simple scientific experiments. Typically they communicated a few basic observations back to earth with low-powered radio transmitters at a fairly slow speed. Communications satellites came quickly.

The United States launched its first satellite, Explorer I, on January 31, 1958 (Fig. 1-1). On December 18 of that year, SCORE (signal communications by orbiting relay equipment) was launched—the biggest object orbited by the United States up to that time. SCORE was more of a propaganda vehicle than anything else. It consisted of part of an Atlas ICBM (intercontinental ballistic missile) with a radio transmitter on board that broadcast a Christmas message recorded by President Dwight Eisenhower (Fig. 1-2). There was no provision for changing the message after the satellite was launched, but it was a true communications satellite.

During these early years there were three basic approaches being considered to the problems of communication via satellite. The satellite could actively receive and retransmit messages, passively reflect messages back to earth, or it could receive and record them for retransmission on command. The United States experimented with all these methods in the first five years.

Each of the methods had advantages and disadvantages. The passive reflectors were by far the simplest. They required no on-board power or sophisticated radio gear in space. These were big advantages at a time when payloads were measured in pounds, and a photovoltaic power supply for a satellite cost about $100,000 a watt, exclusive of launch costs.

The first of the reflector satellites to go into space was Echo, launched in August, 1960. The satellite consisted of a 100-foot aluminized plastic balloon that was inflated once it reached orbit. Echo was certainly the most spectacular of the early satellites for observers on earth. The aluminized skin reflected sunlight and radio waves, making it brighter than most of the stars in the night sky. Because Echo was in a

Fig. 1-1. America got its start in space with the Explorer satellite launched by a Jupiter-C on January 31, 1958. Except in the most technical sense, Explorer was not a communications satellite, but thousands of amateur radio operators listened to its signals (courtesy NASA).

Fig. 1-2. The Atlas Centaur was one of the workhorses of the communications satellite program (courtesy NASA).

Fig. 1-3. The Echo II goes into orbit on a Thor Agena B rocket. Echo was easily visible from the ground during its lifetime (courtesy NASA).

low-earth orbit, it moved quickly across the sky and was a highly visible (Fig. 1-3).

Another early experiment eliminated the satellite altogether. Instead, the rocket released a cloud of metal chaff not unlike that used by the military to defeat radar detection. The metal whiskers were cut to length, and each one was a resonant half-wave dipole. After the whiskers stabilized in a doughnut-shaped cloud around the earth, radio signals could be bounced off the chaff with each whisker serving as a reflecting antenna.

Although passive reflector satellites were popular in an era when satellite transmitter outputs were measured in milliwatts, there were some serious drawbacks to the concept. While the orbiting component of the system was very simple, the total communications system was not. The signals were weak and distorted by the time they reached their destination. They required both a powerful transmitter and receiver to make them operate effectively. The balloon satellites like Echo also had to be located precisely to beam a signal at them and they tended to lose most of the signal to backscattering (Fig. 1-4). The chaff reflectors were highly frequency-specific. Because the individual fibers were randomly oriented, the signal was returned along multiple paths. Multipath time delay and phase incoherency, among other things, made the signal hard to unscramble.

On October 4, 1960, just a few months after Echo, the United States Army Signal Corps launched Courier, the first active repeater satellite. In addition to a transmitter and receiver, Courier had facilities for storing messages and retransmitting them later. Like all its predecessors, Courier was in a low-earth orbit. That meant it was not in range of all its ground stations at any one time.

That was fine for the army, but it was not ideal for a civilian communications satellite. The idea of communications satellites was attracting serious attention from the companies responsible for maintaining long-distance communication. These companies were facing a communication explosion that threatened to overwhelm them.

In 1950 a total of 1.5 million trans-Atlantic telephone

Fig. 1-4. The Echo satellites were nothing but large metallized balloons off of which signals were bounced. This was one of several space communication techniques that were discarded in favor of geosynchronous satellites (courtesy NASA).

calls were made. By 1960 there were 3 million over-the-Atlantic phone calls, an the existing cables were in danger of overload. When planners at American Telephone and Telegraph (AT&T) looked ahead to 1970, they saw 21 million calls being made. Either the growth of an extremely profitable segment of the telephone business would have to be artificially stunted, the bottom of the Atlantic would have to be paved with telephone cables, or another way of transmitting messages would have to be found.

By 1960 satellites with all their imperfections, offered numerous advantages overlaying more cables. Unlike cables, a communications satellite systems would be fairly easy to expand. It only cost about twice as much to build a terminal capable of handling 100 voice channels as it did to build one to handle 10 channels. The satellites were expensive to build and launch, and they would probably only last about two years. The cost would not rise significantly as the number of channels increased. The same network of ground stations could service many different satellites. The calculations indicated that a satellite system capable of carrying 1,000 telephone channels with 30 ground stations on both sides of the Atlantic would cost about $100 million to build and another $20 million or so a year to maintain. It would have a revenue potential of about $100 million a year.

COMSATS GO COMMERCIAL

On July 10, 1962, Telstar I, the first commercial Comsat, was launched by the National Aeronautics and Space Administration (NASA) for AT&T (Fig. 1-5). Telstar I was unimpressive compared to what would come later. The satellite had a 2.25-watt transmitter operating at about 4 gigahertz (GHz) and a receiver tuned to 6 GHz. The ground stations used 2-kW transmitters and special horn-reflector receiving antennas with maser amplifiers to give them very high gain with low noise.

From our point of view, there were two particularly significant things about Telstar. One was the orbit chosen. The satellite was launched at a 44-degree inclination to the equator at a highly elliptical orbit for a satellite of that time. The *perigee* (closest point to earth) was 950 kilometers high, and the *apogee* (point furthest from earth) was kilomeers up. Telstar II, launched in May of 1963, had an even more elliptical orbit with a perigee of 973 kilometers and an apogee of 10,783 kilometers.

This was anything but an accident. The main purpose of the Telstar Series was to provide communications between

Europe and the United States. *Geosynchronous orbits*—orbits in which the satellites' position relative to the earth never changes—were still out of reach. The orbit chosen gave maximum time in direct communication with both sides of the Atlantic.

One of the basic laws of orbital mechanics states that orbiting bodies sweep equal areas of their orbit planes in equal times. A satellite near the high point of its orbit seems

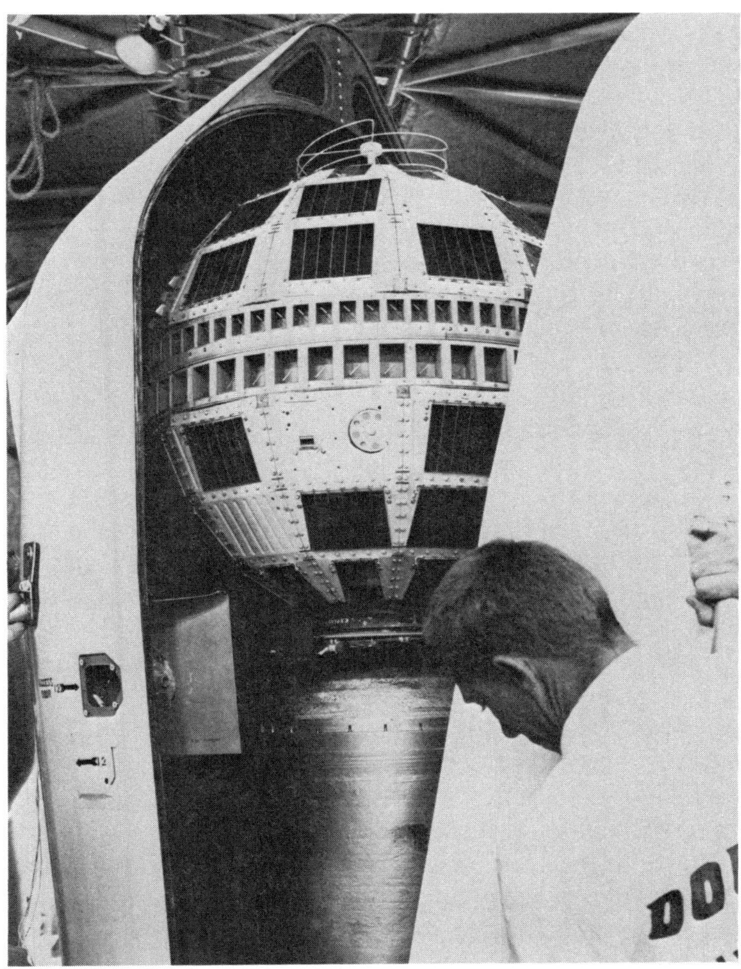

Fig. 1-5. Telstar was the first communications satellite to capture the public's imagination (courtesy NASA).

to move more slowly through the sky than at the low point. That is because a relatively small angular motion at a large distance sweeps out as much area as a larger angle at a shorter distance. By using this principle, the engineers on the Telstar project were able to design an orbit that gave the satellite a longer time in the sky's most useful part.

This wasn't perfect. The earth still tended to rotate under the satellite, so it was only in the best possible position part of the time. The flow of messages between Europe and the United States isn't constant. Messages are more numerous during business hours on both continents. Again, Telstar orbits were designed to give the maximum amount of time over the Atlantic in this window of a few hours.

The other significant thing about Telstar was its capacity to carry communications. It could handle 600 telephone channels simultaneously—or one television channel.

In the early 1960s nobody was thinking in terms of transmitting television via satellite. Clarke had mentioned the possibility back in 1945, and there had been other discussions since. The engineers working on Comsats, though, realized that the bandwidth necessary to transmit a television signal would strain the capacities of existing or planned satellites.

Television signals would have to be transmitted real-time. That meant a satellite in a more or less circular low-earth orbit would only be in a range of the transmitting and receiving stations simultaneously for a few minutes at a time. Transmitting a television picture via satellite would be very expensive because it would tie up the equivalent of hundreds of telephone lines at once.

One of the reasons the idea was attractive in spite of the expense and difficulties was the lack of alternatives. The bandwidth requirements made it impossible to transmit television via undersea cable, and signals at microwave frequency couldn't be bounced off the ionosphere. Television was becoming an increasingly important part of American life, and television news was becoming the principal source of information for people. To show people what was happening in

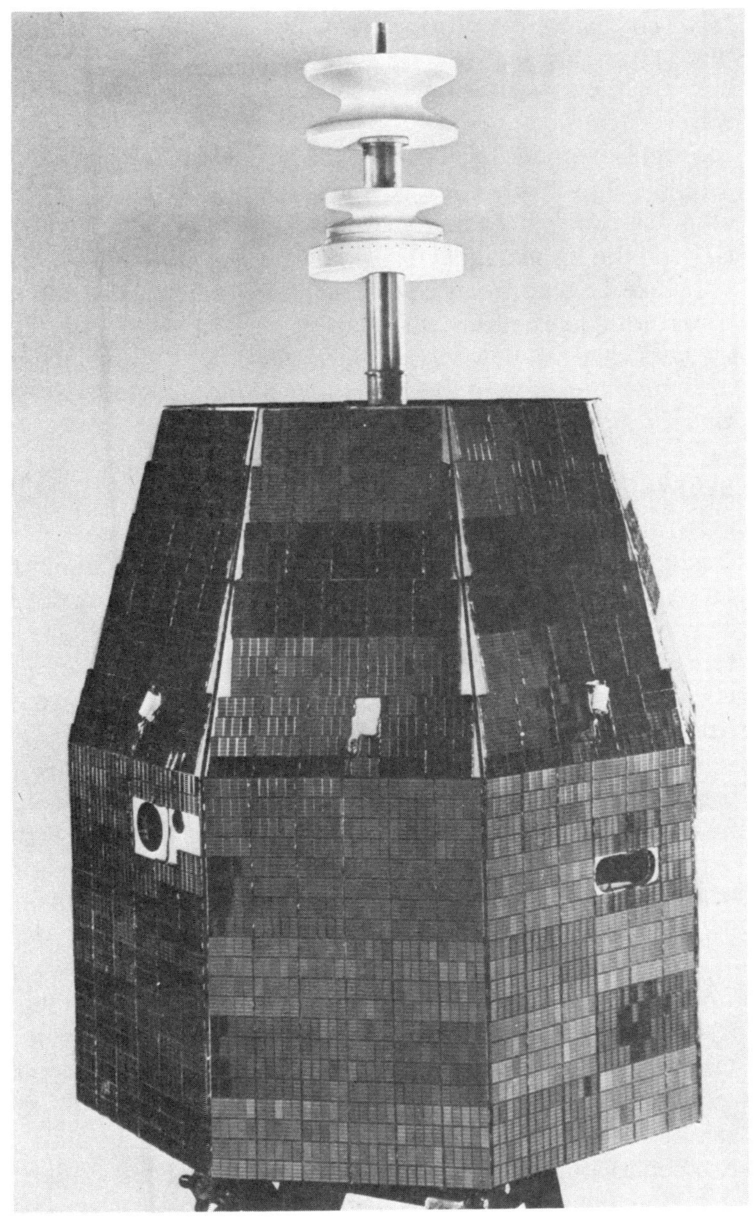

Fig. 1-6. The Relay satellite weighed 169 pounds—less than the first Sputnik (courtesy NASA).

other countries, television networks had to rely on shipping film across the oceans, meaning a delay of hours or days between the time the film was shot and the time it was on the air.

In December of 1962 NASA launched RCA's Relay satellite (Fig. 1-6). Like Telstar, Relay went into a highly elliptical orbit. It had an 11-watt transmitter and used 1.7 GHz on the uplink and 4.1 GHz on the downlink.

Like Telstar, Relay could carry a television signal. Relay transmitted live telecasts of a parade in Moscow to American viewers, and it showed European viewers the ceremonies opening the exhibit of the Mona Lisa at the the National Art Gallery in Washington, D.C.

GEOSYNCHRONOUS SATELLITES ARRIVE

All the early Comsats shared a basic problem: they traced a moving path through the sky. The receiving antennas on earth had to be able to move to cover a good portion of the heavens in order to pick up the signals and Doppler effects in the signals were introduced. This complicated the already complex business of receiving and decoding the weak signals from the satellites.

A worse problem was that it was hard to maintain coverage over a particular part of the globe. A record-and-relay satellite could store messages and transmit them on command to the waiting ground stations, but this was only satisfactory for teletype messages. It wouldn't work for telephone calls or television pictures.

One possible solution would have been to put up a network of satellites in similar orbits. Given multiple satellites, it would be possible to transfer from one just dropping below the horizon to one just rising in the sky and thus maintain continuous contact. Another possibility would be for the satellites to hand the message off to each other until it reached a satellite in view of the intended receiving station.

Both these methods involved complex circuitry to cancel out the Doppler effect. They also required many satellites. Assuming that the satellites were orbiting at an altitude of 4,800 miles, 10 to 40 satellites were required to set up such a system, depending on the elevation angle of the satellites above the horizon and how certain people wanted to be of being able to communicate at any given instant. If the satellites were in a lower orbit, it might take as many as 100 to do the job.

The reason the height of the orbit makes a difference is that the speed with which a satellite moves depends on how close it is to the center of the earth. You can demonstrate the effect by tying a spool to a piece of string and swinging it around your head. If the string is short, you need to swing the spool quickly to make it stand straight out on the string. Let out a little more string, and the spool moves more slowly.

The earliest satellites orbited the earth about once every 90 minutes. A satellite in a circular orbit at a height of 1,000 miles has an orbital period of about two hours. At 12,100 miles the orbit period is about 12 hours, and at 23,500 miles the orbit period is once every 24 hours. At that altitude the satellite keeps pace with the rotation of the earth, and the satellite seems to hang in the sky. This is a geosynchronous orbit, and it is the heart of modern satellite communications.

The first Comsat to occupy a geosynchronous orbit was Syncom I, built by Hughes Aircraft Company and launched on February 14, 1963 (Fig. 1-7). Syncom II, launched in July of that year, successfully achieved orbit. Because it wasn't quite on the equator, Syncom II didn't really stand still in the sky. Instead, it traced out a figure eight motion crossing over the 55-degree West meridian.

In order to get no apparent motion, a satellite needs to be in the plane of the earth's equator. There will be a figure eight motion above or below that plane due to earth's tilt on its axis.

The utility of Comsats for international communications was obvious by this time and the rush was on. To control things, and because international law requires government participation in international space communication, the United States set up the Communications Satellite Corporation in 1962. This quasi-public corporation is owned by the general public and the common carriers who use international Comsats. One-fourth of its board of directors is appointed by the government. The company owns and operates the Ameri-

Fig. 1-7. The Syncom satellite goes into space on board a Delta rocket (courtesy NASA).

Fig. 1-8. Early Bird was launched in 1965 (courtesy NASA).

can part of *Intelsat*, the International Satellite organization satellite system. It also provides certain management services to Intelsat and, in partnership with the common carriers, owns and operates earth stations in the United States, Puerto Rico, and Guam.

Although the law was passed in 1962, the Communications Satellite Corporation wasn't organized until 1963. Intelsat wasn't organized until 1964.

In 1965 Intelsat inaugurated service with its first satellite, Early Bird (Fig. 1-8). On June 28, 1965, former President Eisenhower was interviewed for an international television audience. This was followed with live coverage of a Soviet track meet on July 31, 1965.

Early Bird could handle 240 telephone circuits or one television channel. Each succeeding generation of satellites could handle more channels. Intelsat 2 in 1967 also had a capacity of 240 channels, although of a higher quality. Intelsat 3, launched in 1968, could handle 1,200 channels (Fig. 1-9). Intelsat 4, launched in 1971, had a capacity of 4,000. Intelsat 4a in 1975 could handle 6,000 and Intelsat 5 in 1979 could handle 12,000 (Figs. 1-10 and 1-11).

GROWTH OF DOMESTIC BROADCASTING

The coming of intercontinental television transmission revolutionized communications. An argument can also be made that it revolutionized American politics. Live-by-satellite broadcasts turned the Vietnam War into the "The Living Room War" for Americans. In the 1960s and early 1970s no evening news show was complete without film clips of strafing jets and burning villages from Southeast Asia. The immediacy that satellite transmissions gave the Vietnam War had an impact that was lacking in earlier conflicts.

It wasn't just the war. When American leaders went abroad, television news crews were there to record them. Satellites sent the images back. International diplomacy suddenly has an unprecedented impact on domestic politics as never before. A president or secretary of state knew that his

Fig. 1-9. Intelsat III being launched from Kennedy Space Center (courtesy NASA).

Fig. 1-10. Intelsat V being readied for launch. The satellite can handle 12,000 two-way telephone calls and two color television channels at the same time (courtesy NASA).

Fig. 1-11. Intelsat V was the first of a new series of communications satellites (courtesy NASA).

remarks—and actions would be on the news all over the country, and he had to consider the domestic and international consequences of his actions.

Nevertheless, domestic television still traveled over a network of leased telephone lines and microwave beams. Satellite television was an accepted tool, but only across oceans.

There were several problems. The networks had spent years building up overland communications networks. They had well-proven routes and large investments in facilities. There wasn't anyone else who needed to transmit television programs over long distances in the late 1960s.

The other problem was regulatory. The Federal Communications Commission (FCC) historically has moved slowly on approving new communications technologies. It hasn't been unusual for a decade or more to elapse between the time permission is sought to commercialize a technology and the time the FCC gives the go-ahead. This has been true for everything from television in the 1930s to videotext systems in the 1970s and 1980s.

This is partly bureaucratic conservatism and partly a recognition that once such decisions are made, they cannot be lightly or easily unmade. The frequencies, equipment standards, and other technical specifications laid down in original order will probably be permanent because of the money invested in meeting them.

For example, it's been obvious for years that our present radiotelephone system was a mistake. There are far too few channels available for the number of potential customers. Changing the system has been a long, slow process that isn't over yet.

The third problem is balancing contending forces. Every new communications technology offers advantages to some people and disadvantages to others. Theater owners and radio broadcasters were leery of television. Commercial television stations opposed pay television, and so on. The desires and claims of all these groups have to be weighed and balanced before a decision is made on a new kind of communications technology.

In 1966 Western Union filed an application with the FCC for the first domsat (domestic communications satellite). It wasn't until 1972 that the FCC finally reached a decision. The first of Western Union's Westar satellites wasn't launched until 1974.

Domsats presented an especially tricky problem because in addition to the commercial interests, there was also a very sticky question of control and ownership involved. On one side were the people who felt that the United States should own and control all domestic communications satellites serving the United States. They argued that the government had developed this powerful new technology, and the government must control it.

At the opposite end of the spectrum were the people who believed that, except for assigning frequencies to be used, the government should have no control whatever over domsats. They argued that unlike most regulated forms of communications, such as radio and telephone service, satellite communications weren't natural monopolies. They said that entry into the field would be cheap, and there were potentially many channels available. There was no need to tie satellite communications up with the net of regulations that surrounded many other forms of communications.

In the end the FCC came down more or less on the side of free enterprise. Under the so-called "open-skies policy" announced in 1972, any common carrier (a company providing services to anyone who could pay for them) could launch domsats and compete in the satellite market.

One of the significant things about this decision was that the satellite operators did not have to go back to the FCC and get permission every time they wanted to offer a new service. They could offer any communications satellite service within broad limits. When the operators wanted to start relaying television programs, they didn't have to wait a decade or so for FCC approval.

Because of the regulatory lag, other nations beat the United States into the domsat field. The first country with a geosynchronous domsat was the Soviet Union. In 1965 the Soviets launched their Molniya 1 to provide communications

Fig. 1-12. The Canadian Anik-B satellite (courtesy NASA).

to Siberia and other distant parts of the Soviet Union.

Another country that faced the problem of communicating with a sparse population spread over vast distances of hostile terrain was Canada. In 1972 the Canadians launched (via NASA) their Anik 1 satellite, which was built for Telsat Canada by RCA (Fig. 1-12). In 1973 they followed up with Anik 2.

In 1974 Western Union launched Westar 1 and 2, and the United States entered the Comsat age (Figs. 1-13 and 1-14). In 1975 RCA launched Satcom 1, and in 1976 the company followed up with Satcom 2 (Figs. 1-15 through 1-17).

Both Westar and Satcom satellites use the 6-GHz band for uplinks and rebroadcast on the 4-GHz band. Westar has 12 transponders, each capable of carrying one television channel or 1,200 one-way voice circuits. They transmit on 3,700 to 4,180 MHz with 36-MHz bandwidth per channel. The Satcom satellites use the same frequencies, but they pack 24 channels into the same space by polarizing the alternate channels vertically and horizontally to minimize interference between channels.

Fig. 1-13. Solid-fuel boosters were used to augment the thrust of the Delta vehicle that launched the Westar satellite. The present generation of communications satellites is approaching the limits of conventional launch vehicles (courtesy NASA).

Fig. 1-14. The Westar communications satellite (courtesy NASA).

CABLE TELEVISION GOES SATELLITE

By the time the FCC got around to approving domsats, there was a segment of the television industry that was ready to buy time. Although the networks were slow to show an interest because of their functioning land-based systems, cable television operators were interested from the start.

Cable had started as a way of providing television to remote communities that couldn't get a clear signal with individual antennas. Pay TV services picked it up as a way of getting signals to their customers without having to broadcast

Fig. 1-15. RCA's Satcom satellite lifts off aboard a Delta launch vehicle (courtesy NASA).

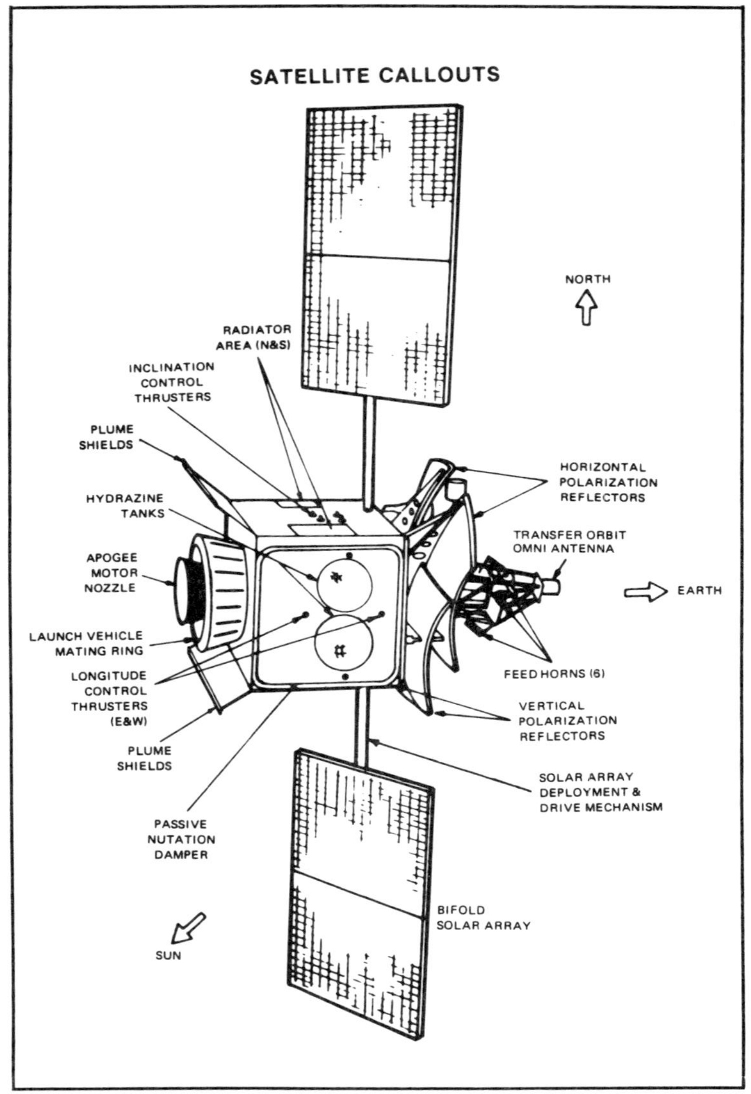

Fig. 1-16. The anatomy of a communications satellite, in this case an RCA Satcom satellite (courtesy RCA Government Systems Division).

them. By 1975 Home Box Office (HBO), the largest of the cable companies, was looking to tie its scattered systems together to offer better quality programming to all its customers simultaneously.

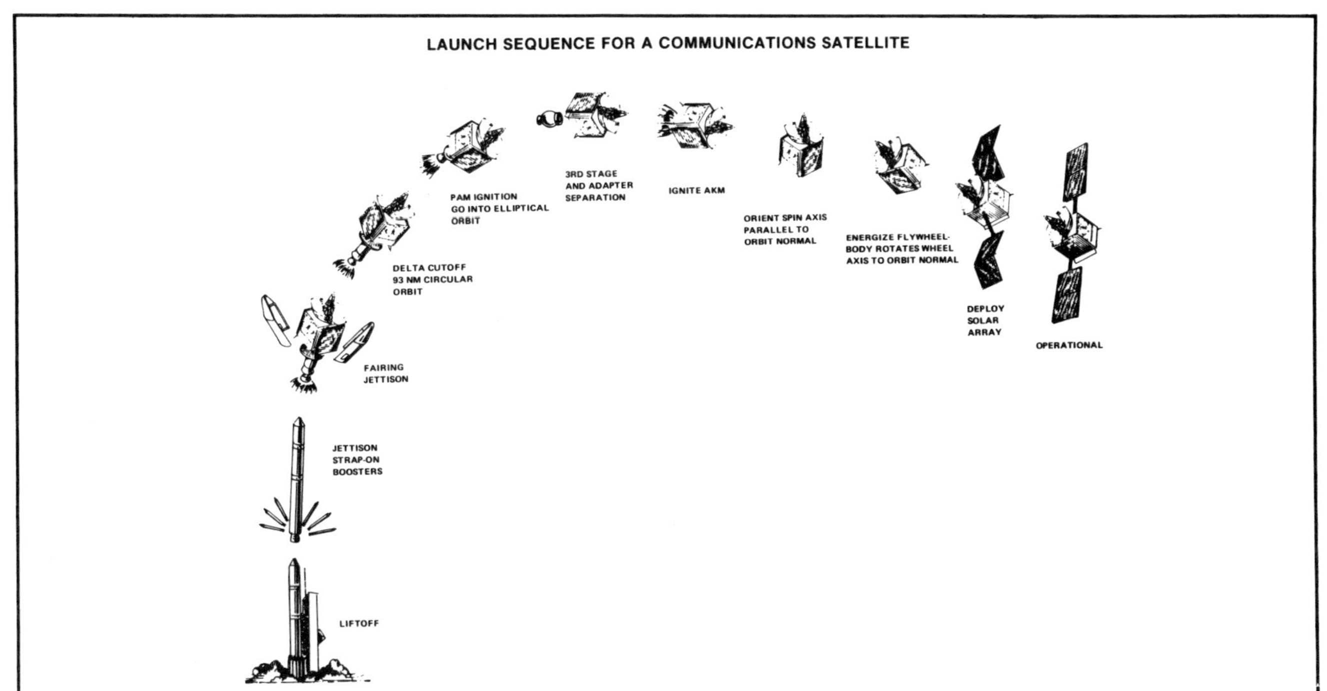

Fig. 1-17. The various stages that an RCA Satcom satellite goes through from lift-off to operation in orbit (courtesy RCA Government Systems Division).

HBO wanted to offer first-run movies and other events to its customers as a way of building business. Satellite transmission to earth stations at each community was the obvious way to go.

HBO rented a transponder on Satcom 2 to provide its service. In December of 1976 Southern Satellite Systems (SSS) became the second cable company to go satellite. It also rented a transponder on Satcom 2 so cable owners could pick up both signals without having to shift their antennas. SSS grew to 150 systems served in the first year. Other cable suppliers got the same idea and also rented space on Satcom 2.

Southern Satellite Systems was the brainchild of Ted Turner, a flamboyant southern entrepeneur who owned WTCG, an independent UHF station in Atlanta. Turner lost about $2 million in the first two years that he owned the station before he changed its programming to movies and sporting events. He saw in satellite television the opportunity to increase his customer base by putting his station's signal on a satellite and offering it to cable television operators all over the south. The first television superstation was born covering an entire region instead of just one city.

The idea of superstations spread rapidly. Signals from WPIX in New York City are transmitted by Eastern Microwave, Inc., to cover most of the northeast. Other stations whose programming is carried by satellite include WOR-TV and WNEW-TV (New York City), WGN (Chicago), and KTVU (Oakland-San Francisco).

One of the things that has helped the spread of superstation programming is the FCC's "wild card" rule. Under FCC regulations a cable operator must black out a syndicated television program being broadcast from another city if a local station is carrying the same program. Because most cable systems carry one or more out-of-city commercial channels, they often have holes to fill. One way to fill those holes economically with high quality programming is to use a movie or sporting event off a superstation.

Other groups that were quick to see the advantages of

satellite transmissions were the new networks springing up in the 1970s. Stations with special interests found they could pool their programming and offer a better quality product to their customers. These new networks weren't as big as the established networks and weren't trying to compete with them directly. They were aimed at more specialized audiences. Some were religious, like the Christian Broadcasting Network; some were ethnic, like SIN, the Spanish Television Network. The biggest of the new networks was the Public Broadcasting Service (PBS)—the network of educational television stations. PBS started broadcasting via satellite in 1978.

With the growth of satellites as common carriers, companies have gone into the business of packaging satellite time. Packagers contract with the satellite owners for large blocks of transponder time and then resell the time by assembling special networks for special events. Among the companies active in this field are Hughes Television Network and Robert Wold Company.

Hughes has a contract with Western Union to buy 5,000 hours a year on Westar 2. It sells 2,900 hours a year to SIN, some time to various small users, and then uses the rest to broadcast cultural and sporting events, generally to national and regional ad hoc networks. This allows television stations to show the away games of their city's hometown sports teams.

PROBLEMS AND THE FUTURE

The biggest single problem facing satellite television today is that it uses a limited resource. There is only an arc of 40 degrees in the sky where satellites can be put to cover the entire United States. Because the signals from the satellites are so weak, they must be spaced 4 to 5 degrees apart to minimize interference. There are about 10 slots available for satellites and only a limited number of transponder positions available on each one. Presently almost all the slots are occupied, and the demand for additional transponders is growing.

There are ways around this dilemma. The way that is chosen will have a major impact on the future of satellite television.

It might be possible to broadcast more than one channel on a single transponder. RCA has done some work with a compression technique to transmit two signals on one transponder. CBS has actually sent experimental broadcasts using the STRAP system developed by CBS Laboratories to combine two channels on one transponder.

The problem with these techniques is that they all run into a fundamental limit imposed by communications theory. You simply need a certain bandwidth to transmit a given amount of information—no matter how cleverly it is encoded. There is some potential for improvement here, but it is limited.

Another possibility is to use a higher frequency for the satellite signals. The amount of bandwidth needed to transmit a given signal in a given scheme is constant, but the available bandwidth increases as the frequency of the signals increases. This is being strongly looked at for the next step in satellite television—broadcasting directly to the home.

There are also some problems to be overcome. The next available band is the Ku band at 12-14 GHz. At these frequencies weather becomes a problem because the wavelength is comparable in size to a raindrop. A signal sent through a rainstorm will lose a significant amount of energy because it is absorbed by the rain. This is why the Ku and the higher-frequency K bands are not used much for terrestrial communications.

A satellite signal travels through a rainstorm vertically or at a fairly steep angle, not horizontally as a signal between two points on earth does. It isn't likely to have to travel more than five miles through the rain. This means that the Ku-band signals would still have enough strength to carry an acceptable signal.

The second problem is cost. The cost of Ku-band receivers is now quite high. Transistors designed to operate at those frequencies cost as much as $300 each. Manufacturers interested in Ku-band broadcasting claim they can bring the price down to as little as $160 for the entire receiver, but this remains to be seen.

Most of the interest in Ku-band satellites presently centers on direct-to-home broadcasting (DBS). We will discuss it further in Chapter 10.

Another approach to the problem of crowded bands is brute force. Use bigger, more powerful transponders or much more sensitive receivers. The basic difficulty is that the satellites can interfere with each other's transmission. If the transmitters were more powerful or the receivers were more

Fig. 1-18. The launching of the space shuttle opened a whole new era in satellite communications (courtesy NASA).

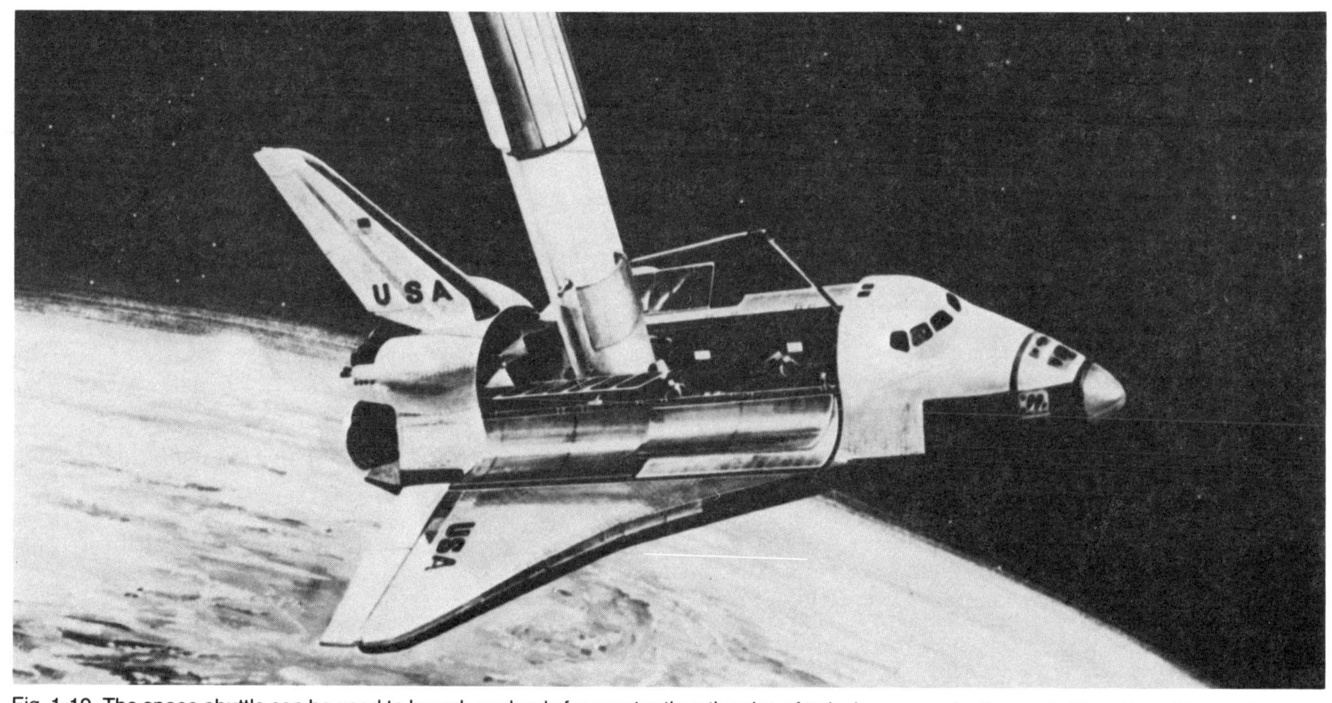

Fig. 1-19. The space shuttle can be used to launch payloads far greater than the size of today's communications satellites. Here it is used to put a large telescope into orbit (courtesy NASA).

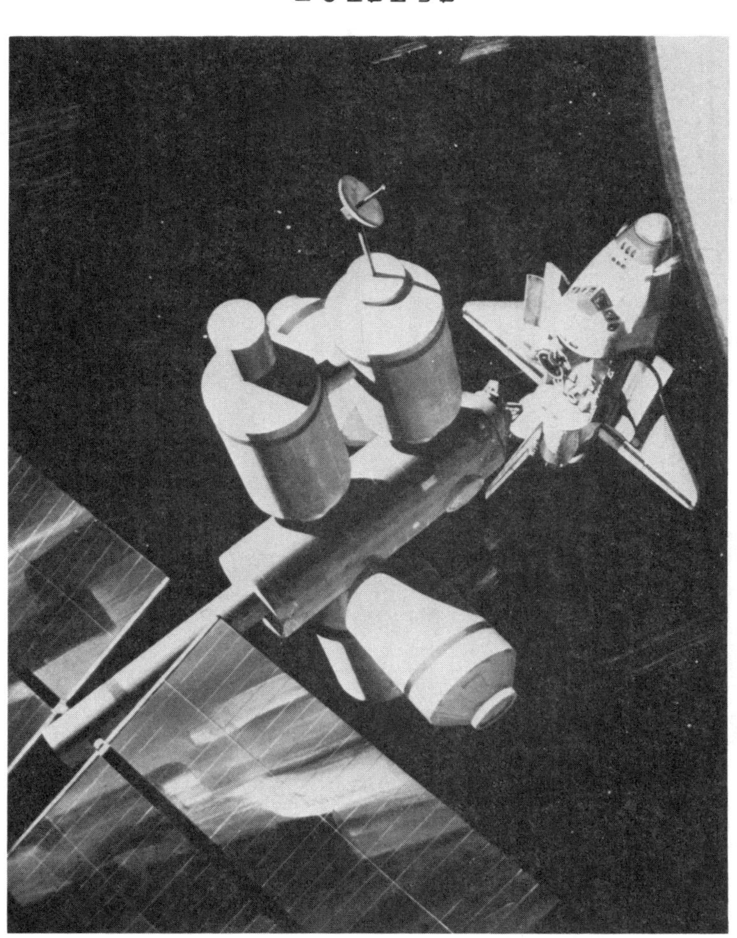

Fig. 1-20. The space shuttle can eventually be used to carry up the parts for entire manned communications satellites that will be assembled in low earth orbit and boosted up to geosynchronous orbit (courtesy NASA).

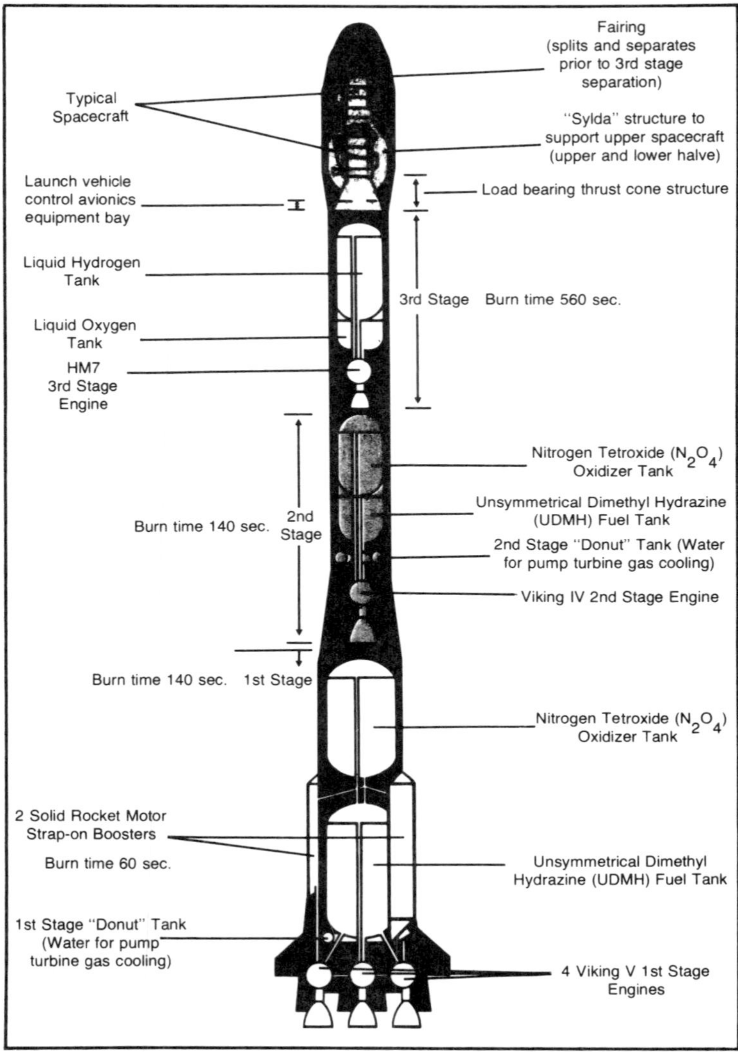

Fig. 1-21. The European Ariane is a competitor for the space shuttle for launching communications satellites (courtesy GTE).

sensitive, it would be possible to pack the satellites closer together in the sky. In our opinion, this is probably the way the problem will be solved in the near term.

Until recently there was a limit on how big a payload could be put in geosynchronous orbit. That limited the size and power of the satellites. Now that the space shuttle is operating, it is possible to loft much heavier payloads up to the geosynchronous orbit (Figs. 1-18 and 1-19). The shuttle can carry the equivalent of a boxcar load into low earth orbit, and boosters can take the satellite from there up into geosynchronous orbit (Fig. 1-20). Note the Ariane 3 in Fig. 1-21.

This has numerous advantages. Bigger satellites can

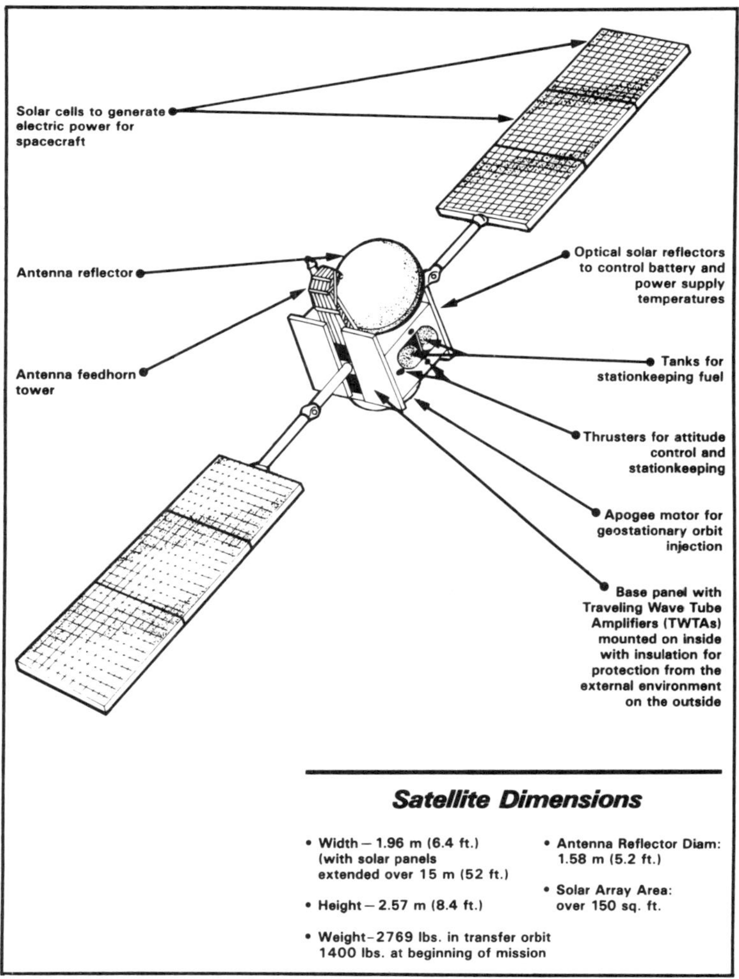

Fig. 1-22. The GSTAR communications satellite from GTE will be used primarily for business communications (courtesy GTE).

carry more transponders and more solar cells to power then. Satellites could weigh 5,000 to 10,000 pounds instead of the 3,000 pounds that is now the maximum. This would allow them to use 30-foot antennas with 5,000 watts of power instead of 8-foot dishes with 1,000 watts now used. Such satellites could be spaced 1 degree or less apart and provide better signals than the present generation.

Eventually we may see huge communications stations that are assembled in low-earth orbit from several shuttle loads and boosted into geosynchronous orbit. They could carry thousands of television channels on a single satellite. This won't come until the end of the century.

In the long run bigger, more powerful transmitters are probably more attractive than increasing receiver sensitivity. The bigger satellites would be expensive, but increasing the sensitivity of thousands of ground stations wouldn't be cheap. The cost of more sensitive receivers rises very rapidly as the gain goes up. To get a major increase (10 times or more), it would probably be necessary to start using masers cooled by liquid nitrogen.

Looking at the impact that satellite television has had on our telecommunications systems and our lives, it is hard to imagine that the entire field is less than 20 years old. It is difficult to know how much greater impact satellite television will have in the coming 20 years.

One thing seems certain. For the rest of our lives, satellites will be in the sky beaming down an incredible variety of information and programs to anyone with the equipment to receive them (Fig. 1-22).

Chapter 2

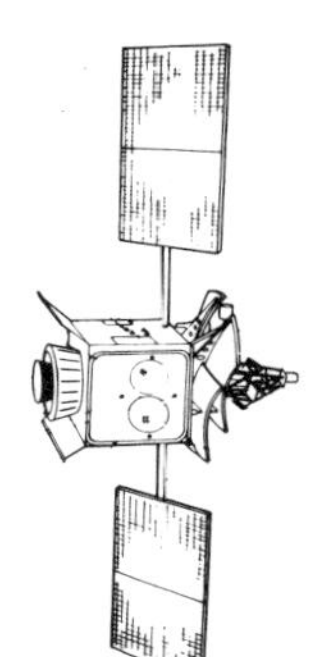

How Satellite Television Works

In some ways the biggest problem with a satellite earth station is that dish out in the backyard. The problem is psychological and not technical. To most people a satellite antenna conjures up images of radar, space probes, and other complicated, high-technology things.

Most people would be shocked to know that what goes on inside an $80 black-and-white portable television is considerably more complex that what happens in an earth station. That big dish and all those little black boxes are actually pretty simple—at least conceptually.

The reason the earth station sells for $6,000 and the television goes for $80 lies in volume production and precision. Televisions are made by the millions, which drives the price down. The production run on a given television set model is often large enough to warrant building an entire factory just to make that one model. In satellite television a production run of 100 is considered large, and one of 1000 is huge. With those numbers you can't justify the types of equipment that cut costs so drastically on televisions. Many home earth station components are virtually hand-built by skilled technicians. That's not cheap at all.

Presumably, this will change as production of earth sta-

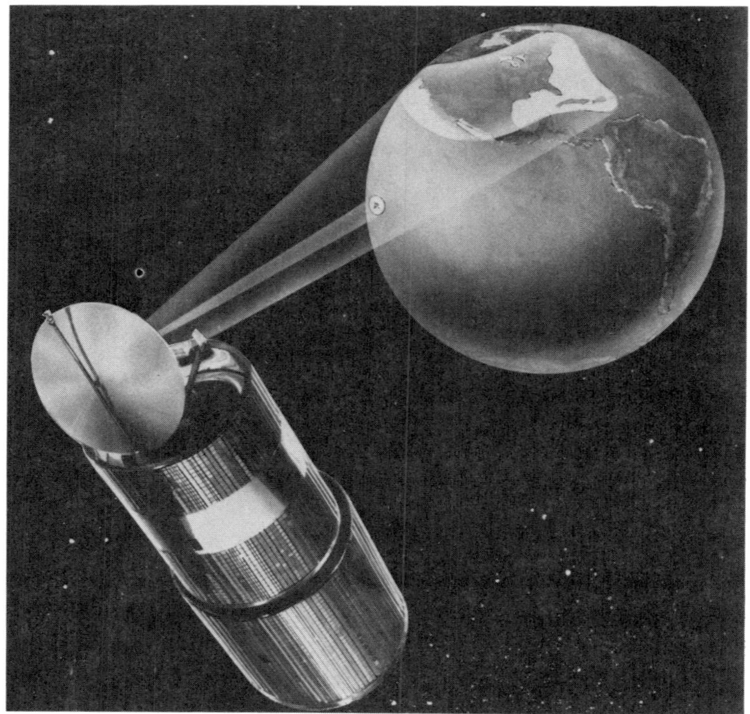

Fig. 2-1. The Westar IV can beam signals to the continental United States and Hawaii (courtesy Hughes Aircraft Company).

tions increases. Rising volume means falling costs, and price tends to track cost fairly closely in the electronics industry. Another possibility is that new technologies will help cut costs. Already microwave integrated circuits are coming into use in other fields. If the innards of an earth station can ever be reduced to a handful of chips, the costs should drop radically.

The need for precision will always be there to keep the cost of a home earth station above that of a television receiver. Because of the signal's frequency, satellite television equipment must be built carefully and to very high standards. Things that wouldn't bother a television circuit will stop an earth station cold.

The process of receiving a satellite signal on your home television starts with the satellite or, more precisely, with one of the transponders on a satellite in geosynchronous orbit. Each transponder is dedicated to a separate communications use. A single bird may have up to 40 transponders on board, with each capable of handling a television channel. With the need for backups in case a transponder fails and other things, a satellite only handles a maximum of 24 television channels (Fig. 2-1).

The satellite receives a signal from a transmitting station down on earth at a frequency of about 6 gigahertz (GHz). The appropriate transponder picks up the signal and beams down a signal carrying the same information at about 4 GHz.

PROBLEMS WITH SATELLITE TELEVISION RECEPTION

There are two central problems with satellite television reception, and both of them concern that signal coming down. The first and biggest problem is the weak signal. A transponder on a modern Comsat has an output of between 5 and 8 watts. That is about the output of a citizens band (CB) radio and perhaps one-third the power of a cheap stereo set. When you're dealing with a signal covering all the United States, a good part of Canada, Mexico, and the Caribbean, this is not much power.

The signal power is so low because the cost in money and ingenuity to put anything into geosynchronous orbit is very large. Weight is crucial in satellites and higher power transponders mean much more weight. It isn't so much that the transponders themselves would weigh more, but they would need a bigger bank of solar cells to generate electricity. This would require a larger and heavier supporting structure. The weight penalty mounts very quickly.

Most of a satellite earth station's design is an attempt to compensate for the signal's weakness. All an earth station really does is pick up that signal, amplify it, and convert it to something your television can use.

The second problem with satellite signals is the frequency. Satellite earth stations must receive signals at frequencies more than 2,000 times higher than the frequencies handled by an ordinary AM radio. Electrical components behave very strangely when you get up in that range. Signals don't flow through wires; they flow along the wires' surface. An ordinary piece of copper wire acts like an inductor (a coil), and a blob of solder can act like a dead short. It is very hard to minimize loss of strength in what is already a very weak signal.

Working at these frequencies takes special techniques and equipment. Regular wire is replaced with waveguides or quality coaxial cable. Meticulous care must be taken in soldering joints and terminating connections. *Alignment*—the process of fiddling with a piece of gear to produce the best possible signal—becomes a major task. This is another reason that home satellite equipment costs so much.

There is no intrinsic reason why you have to use such a high-frequency signal for satellite television. As long as the frequency was high enough that the signal didn't bounce off the atmosphere, you could use any frequency. By the time satellite television was invented, though, the radio-frequency spectrum had been mostly parceled out for other uses. The 4-6-GHz band was what happened to be available. Now let's rejoin our signal, weak though it may be, as it leaves the transponder.

ANTENNA

Most satellites only have two *antennas* to save weight—one to receive signals on and one to transmit them. Because each bird may have 24 or more signals being sent and received at once, the signals are multiplexed by chopping them into small pieces and transmitting a piece of each signal in rotation. This happens so fast the receivers on the ground don't perceive that they are getting their signal in bursts. To prevent interference between adjacent channels, the signal can be vertically or horizontally polarized. This effectively doubles the number of channels an individual satellite can handle. Satellites that don't polarize their signals can only handle half as many channels even if they have the same number of transponders on board.

As the signal approaches earth, it fans out to cover thousands of square miles. Not all satellites have such broad coverage. For the satellites in which we are interested, a broad footprint is an advantage. Some satellites do have their beams shaped to cover certain areas. This is one reason why some satellites can only be received in part of the country.

By the time the signal reaches the earth, it is extremely weak. To receive it, an earth station's antenna must be pointed almost directly at the satellite. Because the satellite has no apparent motion from the earth's surface, it is only necessary to line the antenna up with the satellite once. No elaborate tracking system is needed to keep it on target.

The space-age shape of the antenna acts like the mirror in a telescope to focus the signal. The microwave energy that strikes the antenna surface is reflected back to a single point. Most antennas are parabolic to provide the maximum signal concentration at the focal point. Some antennas are spherical. They don't focus the signal quite as well, but they are easier to refocus on a nearby satellite. They also allow you to receive signals from two satellites at once, if you have two of everything else that it takes to pick up signals.

The antenna acts as the first stage of the amplifying system and gives a gain of 40 dB or more. The bigger the area of the antenna, the more signal it can concentrate and the better its performance will be. This is why some companies sell add-on sections that can be bolted to the periphery of their antennas to increase performance in places with a weak signal and much noise.

FEED HORN AND LOW-NOISE AMPLIFIER

At the focal point of the antenna is the *feed horn* and the *low-noise amplifier.* The feed horn assembly looks like a leftover part of an air conditioning duct. It is hollow, generally rectangular in cross section, and tapers to where it is attached to a nondescript metal box with some cables coming out of it. The box is the low-noise amplifier.

The feed horn picks up the signal that has been concentrated by the antenna and feeds it into the low-noise amplifier. On some earth stations it also distinguishes between signals with different polarization. To go from a horizontally polarized signal to a vertically polarized one, you just rotate the feed horn.

The low-noise amplifier is roughly analogous to the preamplifier in a stereo system. It amplifies the incoming signal with a minimum of distortion and an absolute minimum of added noise, so the other components in the system can work on it.

Added noise is important. It is why a low-noise amplifier (or LNA) is called that, and why it is usually the second most expensive part of the system after the antenna.

Every electronic circuit produces some random noise just from the motion of the electrons in its components. Noise is measured in degrees above absolute zero or degree Kelvin. The lower the temperature, the less built-in noise. A typical home earth station LNA will have a noise rating of 90 to 120 degrees.

The LNA is no more cooled to those temperatures than a 100-horsepower motor has 100 horses inside it. The LNA is actually at whatever the temperature of the air is. Just think of it as a rather mysterious but useful standard of comparison.

CONVERTER, RECEIVER, AND MODULATOR

Things get a bit complicated at this point. There are several ways to do the rest of the job. The functions can be grouped according to one of several schemes, and equipment names vary depending on what scheme is used.

The jobs themselves are pretty straightforward. The amplified signal has to be converted from the 4-GHz range down into something easier to handle. The proper channel has to be selected, and the signal that goes to your television set must be generated. If you have a separate box to convert the signal, you have a *converter* (or a down converter, because it converts the signal to a lower frequency). If you combine the converter with the low-noise amplifier, you have a low-noise converter (LNC). Next comes the *receiver,* which may or may not include the converter and the modulator. The receiver's basic job is to select the signal for the channel you want to watch and reject all others. Finally comes the modulator if it is needed. The *modulator* takes the signal from the receiver and converts it into a frequency a regular television can handle.

Why a separate modulator? Why not just build it into the receiver?

The arrangment is a holdover from the days when all earth station equipment was built for commercial use. You can get a better picture if you are willing to invest in a studio monitor in place of a conventional television set.

A regular television set is designed to accept a signal with a certain bandwidth in a certain range of frequencies. Unfortunately the bandwidth (amount of information a signal carries) for television is rather limited, and that limits the picture's quality. If you can bypass the radio-frequency portion of your television (the parts that pick up the signal from the antenna) and feed it a higher quality signal directly, you can get a much better picture. Computer hobbyists have been doing this for years when they want to use a television set as a monitor with a computer.

MONITOR

Television *monitors* were designed to monitor broadcasts in the studio. The monitors are special sets that don't have the radio frequency circuits built in, and they are capable of producing quality pictures. The component television systems that are just coming into the market also use monitors. In these systems the tuner and audio section are separate, and they can produce "hi-fi" television from a quality signal. A home earth station produces a higher quality signal than a conventional television station. Some people want the option of picking up the satellite receiver's signal before it goes to the modulator.

The satellite television process is no more complex than an AM radio. From the user's standpoint the only big differences between a home earth station and a stereo system are that the earth station components cost more, and you have to be finicky about things like cable runs and connections. Just that—and the big dish out in your backyard.

Chapter 3

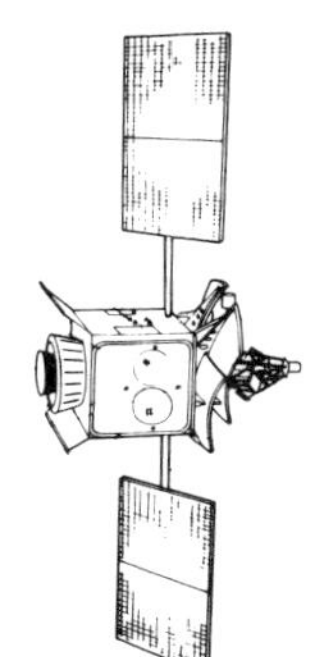

Popular Applications for Satellite Television

If you drive across the country, you'll see big dishes, all pointing south. Antennas are prominent from the Dakotas to Texas and from the barrier islands off the Carolina coast to the apple orchards of Oregon. They are opening a new window on the world for people all over the country.

Satellite television dishes aren't limited to rural areas. They are on the roofs of apartment buildings, in the parking lots of taverns, and in surburban backyards. For urban dwellers the effect isn't as pronounced as it is for their country counterparts, although it is still very real.

A drive through rural America is the real eye-opener. By now nearly everyone is familar with those dish antennas and knows what they mean, though people tend to underestimate their numbers and impact.

The reason is simple: most Americans live in urbanized areas. Americans have been moving out of the cities for more than 20 years, but the suburbs and exurbs are gradually growing together into sprawling megalopolises that cover parts of several states. About 80 percent of the nation's population lives in more-or-less urbanized areas.

One of the characteristics of the places where most people live is good television. This has had a major effect on the growth patterns of home earth stations. They have proven much more popular in rural areas than urban ones. There are more stations per capita in sparsely populated parts of the country than in the densely populated areas. Even in a period when farm income is lower than any time since the depression, home earth stations are spreading like wildfire in farming areas.

The spread of earth stations neatly demonstrates the elementary economic distinction between cost and value. An earth station costs about the same to a farmer in Iowa and a homeowner in the suburbs of Los Angeles. If the Angeleno chooses not to buy one, there are 8 or 10 television broadcast channels to which he can tune. There is the local cable company and possibly a local microwave pay TV system, or he can run down to the local video rental store and rent a video cassette for the evening's entertainment.

The availability of these alternatives limits the utility of a home earth station to the Los Angeles resident. He may like to have one. A home earth station offers him things he cannot get otherwise, but its value to him is not as great as if the alternatives were not present.

For the farmer in Iowa, the alternatives may be a mail-order video cassette club and perhaps one television station that he can hardly pick up with the best antenna he can find. His lack of alternatives increases the value of a home earth station to him, so he is willing to make a larger sacrifice to get it.

TELEVISION AND ISOLATION

The example wasn't chosen to pick on Iowa. There are many areas of Iowa that have very good television. You can substitute Texas, Georgia, or Montana, and the example still holds. It holds in parts of nearly every state.

Many Americans don't watch much television. To them the idea of laying out thousands of dollars just to get television reception sounds slightly preposterous.

Those people may never have lived in a part of the country where there is no television at all. The difference between not much television and no television is qualitative and not just quantitative.

We won't go as far as some critics and claim that television is American culture, but it undeniably plays an important role in our society. Without access to television at all, it is difficult to fully participate in mainstream America.

If most Americans hear a rumor about a crisis or catastrophe, they turn on their television sets to find out what is happening. In those times televisions appear as if by magic in public places to let people follow events.

We got a demonstration of this the day President Reagan was shot. At the time we both worked in the newsroom of a major metropolitan newspaper. The terminals on our desks could access all the major wire services, plus many smaller ones. There were probably 12 of the most authoritative sources of news in the United States available to everyone in the newsroom at the touch of a few keys. The reporters and editors could have compared sources, cross-checked information, and sifted fact (or at least consensus) from rumor at our terminals. For that first hour when no one knew much and wild rumors were surfacing, most of us were glued to the newsroom television set.

Some psychologists say that television is a unifying force for Americans in times of momentous events. We turn to television to find out what is going on and to get in touch with ourselves as a people.

Think about how much the events of the last 20 years are summed up for us by images from the television screen. Remember John F. Kennedy's son saluting his father's casket, Neil Armstrong setting foot on the moon, and Richard Nixon's tearful farewell speech to the White House staff. Television has helped to shape our generation just as radio helped to shape the generations of the 1930s and 1940s.

The sense of communion that television provides isn't limited to times of crisis. Many people use television for that every day. Think of the housewives who leave the television on while they're doing housework. They're not watching it; they may not even be in the same room. But they still want the background noise or the feeling of companionship television gives.

Television is more important to most of us than we realize. Being cut off from television in our society means being isolated to some degree. That helps explain why earth stations are so popular in areas where television reception is poor or nonexistent.

It also helps explain why there is strong opposition in Congress to a law that would curtail or eliminate the right of home earth station owners to receive programming. Home earth stations are important to rural areas, and every congressman with a large rural population in his or her district knows it.

In testimony before a House subcommittee considering legislation that might interfere with home earth stations, Congressman Charlie Rose of North Carolina had this to say:

"Home earth stations allow Americans in rural and remote areas the same program viewing opportunities available to Americans dwelling in more populated areas which are served by cable systems. There are also urban and suburban areas which are unserved by cable or any other source of subscription television service, and again, earth stations are the technology which provides these Americans with a diversity of programming. And in those urban and suburban areas which are served by cable systems or STV or MDS systems, home earth stations provide healthy competition. All this is to say that home earth stations provide a tremendous benefit to Americans everywhere—whether they reside in remote, rural, suburban, or urban areas."

Rose is also the chairman of the ad hoc advisory committee on broadcasting set up by the speaker of the House and the chairman of the House Oversight Committee. We will discuss this matter in some depth later when we cover the legal issues associated with home earth stations. Home satellite television is recognized as an important technology even in the halls of Congress.

THE SATELLITE TELEVISION SMORGASBORD

There are currently about 100 channels available off the

satellites that serve the United States and Canada. The diversity of programming beamed down to earth is incredible, and most of it is simply unavailable to anyone without a home earth station.

No city offers all the available services, and it's unlikely that one ever will. You have the services at your fingertips with a home earth station.

No matter how great a choice of programming you have without a home earth station, satellite television offers you more. That has been the trend in television for several years. As cable television and microwave distribution systems (MDS—cable television without the cable) have spread, there has been a proliferation of "extra" channels. You can get more than three channels of television in nearly every city in the United States. Home satellite television accentuates the trend by greatly increasing the number of channels over even the best cable systems.

To many people familiar with television as it is, that's not something to cheer about. More than 20 years have passed since Newton Minnow, then chairman of the FCC, called television a "great wasteland." Little has happened in that time to change things. To most people more than 12 years old, "Gilligan's Island" is the epitome of dumb television. Yet the show is still in syndication and is probably playing in most major markets in the United States today.

Shows like that were not what the pioneers of television had in mind. They envisioned television as a way to bring the best of the world into our living rooms—concerts, plays, and great events as they happened.

Sometimes television brings us those things, but mostly it doesn't. As television sets sold by the millions, television became the new mass medium. It replaced magazines like *Life* and *Saturday Evening Post* as the way for advertisers to reach the maximum number of people at the minimum cost per person. By the 1960s television had driven the mass circulation magazines out of business, with the exception of the news magazines. A spot in prime time gave an advertiser the most viewers for his dollar.

The inevitable result under the American system of

sponsor-supported television was that numbers became all-important. Ratings evolved from the minor marketing tool they had been in the days of radio into the pole star around which programming revolved. The bigger the audience, the more successful the show, and never mind quality, merit, or anything else.

If most Americans want to watch mindless pap, it ought to be their privilege to do so. The problem is that Gresham's law applies as fully to television programming as it does to money. Bad, popular programming tends to drive out good, less popular programming. This is proven nearly every television season when at least one quality series is canceled in six weeks or less.

None of these observations are original. They have been made repeatedly by critics since the 1950s. If the critics have a bend toward social as well as television criticism, they are likely to suggest a solution that involves cutting down on the mindless pap by law, rule, or fiat. That was Minnow's approach to television, and it was behind the FCC's decision several years ago to give an hour of prime time back to the local stations. In theory that local hour was to go for quality programming. In practice it proved a bonanza for the syndicators who sold the local stations previous years' mindless pap to fill the time.

Fundamentally this was the wrong approach. Some people like mindlessness all the time, and most of us enjoy it at least some of the time. The difficulty isn't that some shows cater to a mass audience; it is that almost all of them do. The problem, in other words, is lack of choice.

Even cable, which in theory can solve the problem completely, has fallen victim to the mass-audience view of television. Most cable systems will offer a movie channel, sports, specials, and a mixed bag of feeds from stations in other cities. The cable viewers are stuck with whatever selection the cable company decides to hand them. That is better than what you can get on commercial stations, but it is far short of what is available off the satellites.

Glance through a copy of SatGuide or any other programming magazine and you'll see that there is a far richer world up there than any cable operator offers. As the use of home earth stations expands and broadcasters and advertisers come to understand their implications, you can expect to see even more diversity.

Part of the poverty of network television is that the audience is gerrymandered by geography. Networks and cable operators provide their service to a geographical area, and they have to take as their audience the viewers in that area. The audience for concerts of light classical music might be quite respectable if you looked at all the viewers in the country. If that audience is fragmented in the different station markets, it may be ignored. Collectively it is significant, but the networks and cable operators can't deal collectively. They have to treat each market as a separate entity.

A satellite feed can reach the entire nation at once. Where a network or cable system must divide its audience along geographical lines, a satellite system doesn't. It can lump together audiences that other forms of television fragment. This makes it practical for a satellite system to cater to groups that regular television ignores, and it gives advertisers the opportunity to reach special audiences. Satellite television doesn't just allow diversity of programming; it encourages it.

Presently none of the satellite broadcasters are considering home earth station owners as a separate market. They are aiming for the cable systems. There is no reason why the broadcasters couldn't direct their programming at the home earth station user. This is particularly true of the commercially supported broadcasters who get most of their revenue from advertisers rather than from cable companies' fees.

TUNING IN THE WORLD

Not all that diverse programming represents things you were supposed to see. Some of it is more or less accidentally available.

The classic examples of this are the network prefeeds—the networks' practice of transmitting some of their news and entertainment programs via satellite hours or days before they go on the air. For example, Johnny Carson's

"Tonight" show is actually taped several hours before it airs. It is then sent to the NBC broadcast center via satellite to be shown at the proper time. Shows are also sent to Alaska via satellite in advance of their air time. All of these programs can be received by a home earth station, so you can go to bed early and still catch Johnny's monologue.

There are also private networks that were never intended for public consumption. NASA has a leased channel on Satcom F-2 to send information between the Johnson Space Center in Houston, the Kennedy Space Center in Florida, and Washington D.C. If you tune in on this channel during a space shuttle mission, you may get to see the unedited video from the shuttle before it is released to the networks. See Fig. 3-1.

If you are technically inclined, there are a number of other interesting services you can pick off the birds. The subcarriers surrounding the television channels are widely used to transmit information that isn't related to the programming. For example, AP, UPI, and Reuters all use subcarriers to transmit teletype service to newspapers and television stations around the country. Special equipment is needed to get these signals, notably a receiver capable of stripping out the subcarrier and a terminal (or computer with the appropriate interface and program) to watch it. You can get the news as fast as your local television station off the satellites.

Sometimes you can get it faster. News stories are usually written by the local wire service bureaus and transmitted to a hub point for editing before being sent out on the wire. You can pick up the stories as they go to the hub via satellite.

If you want to get really exotic, you can even tune in on the Russian news broadcasts. Because so much of their country lies far north, the Russians do not use geosynchronous orbits for their domsats. Their network of Molynia satellites travels in highly elliptical orbits much like the ones used by the early American Comsats. If you live in the northern states, Alaska, or Canada, this puts them above the horizon when they are broadcasting.

Picking up the signals takes more than pointing your

Fig. 3-1. Leasat from Hughes Aircraft is specially designed to take advantage of the space shuttle's larger cargo bay and greater payload capacity. It will handle communications for the United States Navy and other military services (courtesy Hughes Aircraft Company).

antenna north. Because the satellites are not in geosynchronous orbit, they have apparent motion from the earth. That means you need a motorized antenna capable of tracking them accurately, and you must know what their relative motions will be. It is possible, though.

If you live on the east coast, you can also pick up the Intersputnik geosynchronous satellite over the Atlantic. This bird transmits between the Soviet Union, Cuba, and several African nations with large Russian presence.

Finding and viewing these "secret" channels is a challenge for home earth station owners. They are not listed in the satellite program guides; you have to dig them out yourself. Many of them change rapidly and some of them, like the teletype channels, require special equipment to receive. They can provide a fascinating glimpse of the world that you couldn't get otherwise.

EDUCATION: FORMAL AND OTHERWISE

One of the reasons that television never lived up to the hopes of its inventors was that the channels rapidly became a scarce resource. Before uhf, you couldn't have more than four or five channels in any city. That limited the number of hours of programming you could show and, combined with the geographical limitations we discussed earlier, meant some things just didn't make it on the air. Education was one.

In theory television is an ideal substitute for the large lecture classes that dominate undergraduate teaching at the college level. Some colleges make use of television by videotaping classes and letting students watch the tapes as their schedule allows. The use of broadcast classes has been very limited.

Even the public television stations don't broadcast many classes. They may have a class in beginning psychology or western civilization at 6:30 A.M.,but few of them do much more.

This is extremely odd in one sense. We are living in a time when there is a need for continuing education as never before. Knowledge is expanding so rapidly and new techniques are being developed so quickly that most people's education becomes obsolete in their lifetimes. Doctors, engineers, teachers, lawyers, and machinists all need to update their educations regularly. Telecourses would let us do that at home, but we don't use television that way.

The reason is that we can't do it with our present setup. There may be 10,000 people interested in taking a course in introductory robotics, but if they are scattered 200 in one city and 500 in another, there is no way that local television stations can serve them effectively.

If those 10,000 people can be treated as a single, national audience, it becomes worthwhile to talk about televising the course. The students would need access to earth stations, but there are more people who meet that condition all the time. Direct broadcast satellite (DBS) systems would be another possible channel for this kind of programming.

A major university such as UCLA or the University of Chicago could agree to sponsor the telecourse. The university would offer credit for the class to those who wanted it and would prepare the course material. The students would register by mail and pay a fee large enough to make the operation self-supporting. It wouldn't have to be a very large fee. If your average class enrollment is 1,000, you don't need much money from each student to pay for the transponder time and production expenses. Exams would of necessity be open book, and students who got a passing grade would receive credit accepted by their local universities.

This is much cheaper than attempting to mail the video cassettes of the lectures—something that has been tried with limited success. The cost of purchasing and preparing the cassettes is so high that the course fees must be prohibitive to cover them.

Not all our education comes in the form of college courses. "Sesame Street" showed another way that television could be used to educate people.

Television is a very powerful medium with an inherent ability to fascinate. If its characteristics are exploited properly, it can be an effective provider of nonformal education.

This is already being done by the educational channels to an extent. Programs like "Sesame Street" and "The Electric Company" help children learn in a way that is almost subliminal. Again, except for the lack of a way to reach the national audience as a unit, there is no reason why such programs couldn't be done for adults and specialized audiences.

CHOICE, DIVERSITY, AND UNIVERSALITY

The applications for satellite television are limited only by our imaginations. Experience shows us that our imaginations are very fertile.

The first popular application for home satellite television has been to provide service in areas that otherwise would have poor or no television. Thousands of people have discovered that a home earth station gives them a far greater choice of fare than they could otherwise receive. Finally, as the home earth station owners are recognized as a separate and truly national television market, there will be more programming aimed at them specifically. That is when the true potential of satellite television will be realized.

Chapter 4

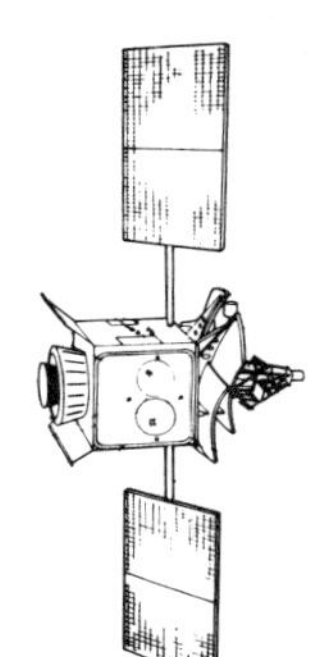

Guide to Satellite Television Programming

The variety of programming available via a home earth station is amazing to someone used to network television. Not only is the variety of the programming greater, but the quality (however you define it) is much better, too. If you're an old movie buff, you'll find that instead of being able to catch a favorite flick once every couple of weeks, you'll be able to find a movie on one of the transponders virtually every night.

The best way to find out what is available is to get a copy of one of the program magazines such as *SatGuide.* Write to Commtek Publishing Company, Box 1048 Hailey, ID 83333. Subscription is $48 a year in the United States. Single copies are available from most earth station dealers. These magazines usually list the programs a la *TV Guide* as well as information on satellite locations and transponder frequencies.

A listing of the programming being carried by the satellites at the time of writing follows. This will give you some idea of what you can expect from a home earth station. As more companies enter the business and more satellites are launched, the diversity of programming will increase.

For the most part we have compiled this list by services rather than by networks or satellite transponders. There really aren't any networks on the satellites—only programming services. Some of the services broadcast 24 hours a day. Others are only on the air for a few hours a week.

We haven't given a time schedule because that can change from month to month. It would probably be out of date by the time the book is published.

Most of these services are for cable operators and not home earth station owners. The fact that a service is listed here in no way implies that the operators intend that it be available to home earth stations. Some have no objection to private receivers; others do. See Chapter 11 for more details.

American Educational Television Network
2172 Dupont Drive, Suite 7
Irvine, CA 92715

Description of Services: programming aimed at licensed professionals who need continuing education.

Satellite & Transponder: Satcom I, transponder 16

Operating Since: January, 1981

Services Available to: cable systems

Policy on Unauthorized Reception: N/A

Plans for Projection of Signal: N/A

Notes on Audio: N/A

Appalachian Community Service Network
1200 New Hampshire Avenue NW #240
Washington, DC 20036

Description of Services: educational and community service programming for the Appalachian region. Emphasis is on programming for professional development, college credit, and continuing education programs.

Satellite & Transponder: Satcom I, transponder 16

Operating Since: August, 1979

Services Available to: N/A

Policy on Unauthorized Reception: N/A

Plans for Protection of Signal: N/A

Notes on Audio: N/A

Black Entertainment Television
3222 N Street, NW—#300
Washington, DC 20007
(202) 337-5620

Description of Services: BET provides music, movies, sports, family, public affairs, and women's programming directed to the black cable subscriber, but attractive to all audiences.

Satellite & Transponder: Westar V, transponder 12X

Operating Since: January 25, 1980

Services Available to: CATV firms only

Policy on Unauthorized Reception: did not respond

Plans for Protection of Signal: did not respond

Notes on Audio: monaural only

Bravo
Rainbow Programming Service
100 Crossways Park West
Woodbury, NY 11797

Description of Services: international performing arts emphasizing film, dance, and music.

Satellite & Transponder: Comstar D-2, transponder 3H

Operating Since: March, 1980

Services Available to: cable television

Policy on Unauthorized Reception: N/A

Plans for Protection of Signal: N/A

Notes on Audio: N/A

Cable News Network
1050 Techwood Drive NW
Atlanta, GA 30318

Description of Services: a news service carrying special interest series in addition to newscasts. A second CNN service is planned for 1982.

Satellite & Transponder: Satcom I, transponder 14

Operating Since: June, 1980

Services Available to: CATV

Policy on Unauthorized Reception: N/A

Plans for Protection of Signal: N/A

Notes on Audio: N/A

C-SPAN (Cable Satellite Public Affairs Network)
400 North Capitol Street, NW
Washington, DC 20001

Description of Services: provides live coverage of the debates and proceedings of the U.S. House of Representatives and special features.

Satellite & Transponder: Satcom I, transponder 9

Operating Since: 1/79

Services Available to: CATV

Policy on Unauthorized Reception: N/A

Plans for Protection of Signal: N/A

Notes on Audio: N/A

CableText
P.O. Box 45684
Tulsa, OK 74714

Description of Services: CableText serves as the carrier for teletext services delievered to both cable head-ends and private users. The services of UPI, Reuters, and Dow Jones are delivered to cable systems. Quotrader is a commodity service for those interested or involved in the commodities business. Multichannel NewsText will start up in late 1982 and is a cable industry newsletter. KEYFAX National Teletext Magazine is a consumer-oriented service.

Satellite & Transponder: Satcom III, transponder 6

Operating Since: October 1980

Services Available to: CATV, except for Quotrader, which is marketed by the originator to private users who do not have cable.

Policy on Unauthorized Reception: "since these services are for cable and the cable operator pays, we will prosecute an unauthorized use of a CableText signal."

Plans for Protection of Signal: none announced.

Notes on Audio: no audio involved, because CableText is carried on the satellite signal's vertical blanking interval.

CBN (Christian Broadcasting Network)
1000 Centerville Turnpike
Virginia Beach, VA 23463

Description of Services: religious-oriented programming including soap operas, morning programs, and college football.

Satellite & Transponder: Satcom I, transponder 8

Operating Since: April, 1977

Services Available to: cable

Policy on Unauthorized Reception: N/A

Plans for Protection of Signal: N/A

Notes on Audio: N/A

Christian Media Network
Box 20121
Bloomington, MN 55420

Description of Services: religious programming including films, family entertainment, and music.

Satellite & Transponder: Satcom I, transponder 16

Operating Since: June, 1981

Services Available to: CATV

Policy on Unauthorized Reception: N/A

Plans for Protection of Signals: N/A

Notes on Audio: N/A

Consat Network
2880 West Oakland Park Boulevard
Ft. Lauderdale, FL 33111

Description of Services: live coverage of musical concerts.

Satellite & Transponder: Comstar D-2, transponder 22

Operating Since: December, 1981

Services Available to: pay-per-view via cable systems.

Policy on Unauthorized Reception: N/A

Plans for Protection of Signal: N/A

Notes on Audio: N/A

Delta
ABC Video Services
1330 Avenue of the Americas
New York, NY 10019

Description of Services: an all-sports network to be a joint venture between ABC and Getty Oil.

Satellite & Transponder: not determined yet

Operating Since: not begun yet

Services Available to: N/A

Policy on Unauthorized Reception: N/A

Plans for Protection of Signal: N/A

Notes on Audio: N/A

Dow Jones Cable News
Box 300
Princeton, NJ 08540

Description of Services: Dow Jones Cable News provides news in 15-minute cycles with emphasis on financial and stock market news. It also features stories from the *Wall Street Journal* and *Barrons*, two of the nation's leading business papers.

Satellite & Transponder: Satcom I, transponder 3

Operating Since: April, 1981

Services Available to: CATV

Policy on Unauthorized Reception: N/A

Plans for Protection of Signal: N/A

Notes on Audio: N/A

Entertainment and Sports Programming Network
ESPN Plaza
Bristol, CT 06010

Description of Services: Twenty-four hour sports programming featuring men's and women's professional and amateur events from national and international sites, including college, Canadian, and USFL football; college and NBA basketball; boxing; Davis Cup tennis; LPGA and PGA golf; auto racing; and studio shows and "SportsCenter," the network's sports news show that airs seven times daily.

Satellite & Transponder: Satcom III-R, transponder 7

Operating Since: September 7, 1979

Services Available to: CATV, MATV

Policy on Unauthorized Reception: "we take all reasonable legal steps justified to prevent piracy."

Plans for Protection of Signal: none announced

Notes on Audio: monaural only

The Entertainment Channel
RCTV
Rockefeller Plaza
New York, NY 10112

Description of Services: Films, cultural programming, children's programming, and BCC programming.

Satellite & Transponder: Westar III, transponder 7

Operating Since: spring, 1982

Services Available to: N/A

Policy on Unauthorized Reception: N/A

Plans for Protection of Signal: N/A

Notes on Audio: N/A

Episcopal Television Network
Box 2060
New York, NY 10163

Description of Services: religious programming aimed at Episcopalians and other Protestant sects.

Satellite & Transponder: Satcom I, transponder 16

Operating Since: September, 1979

Services Available to: CATV

Policy on Unauthorized Reception: N/A

Plans for Protection of Signal: N/A

Notes on Audio: N/A

Eros
Broadcast Programming Inc.
2 Lincoln Square
New York, NY 10033

Description of Services: adult movies—both foreign and American.

Satellite & Transponder: Westar III, transponders 7 and 12

Opeating Since: January, 1982.

Services Available to: N/A

Policy on Unauthorized Reception: N/A

Plans for Protection of Signal: N/A

Notes on Audio: N/A

Escapade
Rainbow Programming Service
100 Crossways Park West
Woodbury, NY 11797

Description of Services: R-rated films and specials as well as entertainment programs from *Playboy*.

Satellite & Transponder: Comstar D-2, transponder 4V

Operating Since: December, 1980

Services Available to: CATV

Policy on Unauthorized Reception: N/A

Plans for Protection of Signal: N/A

Notes on Audio: N/A

Eternal Word Television Network
5817 Old Leeds Road
Birmingham, AL 35210

Description of Services: Eternal Word Television Network is devoted to Catholic religious programming and family entertainment. It broadcasts four hours nightly, seven days a week. There are no fund solicitations on the network.

Satellite & Transponder: Satcom III-R, transponder 18

Operating since: August 15, 1981

Services Available to: CATV and SMATV

Policy on Unauthorized reception: allows private earth stations to receive signal and requests that the network be notified in writing.

Plans for Protection of Signal: not applicable

Notes on Audio: mono only

Gala Vision
250 Park Avenue
New York, NY 10177

Description of Services: GalaVision is a Spanish language premium pay television service offered exclusively on cable. It offers programming 12 hours a day Monday through Friday beginning at 1 P.M. PTS., and 24 hours a day on Saturday and Sunday. It shows first-run movies; novelas (edited especially for GalaVision into 90-minute chapters for viewer convenience); sporting events including boxing, soccer, jai alai, baseball, and bullfighting; and specials.

Satellite & Transponder: Westar IV, transponder 12X

Operating Since: October 26, 1979

Services Available to: CATV

Policy on Unauthorized Reception: not stated
Plans for Protection of Signal: Not stated

Notes on Audio: N/A

Hearst/ABC Video Services
555 Fifth Avenue
New York, NY 10017

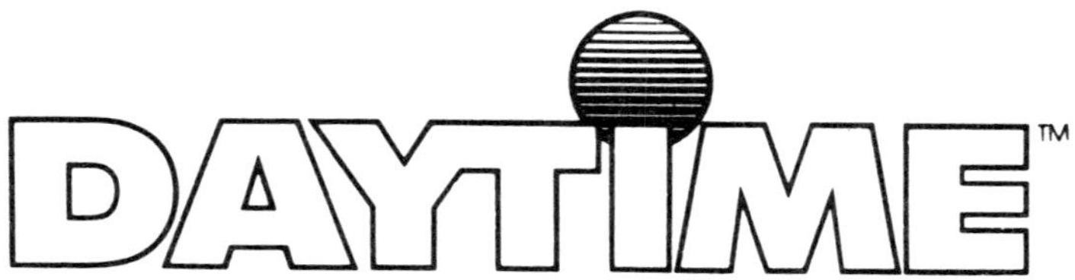

Description of Services: a joint venture of the Hearst Corporation and the American Broadcasting Companies. Hearst ABC/Video provides cultural offerings through ARTS (Alpha Repertory Television Service) and women's programming on its DAYTIME service. The latter is a four-hour-a-day magazine-style show for women five days a week.

Satellite & Transponder: they are offered as part of Cablenet I.

Operating Since: April, 1981

Services Available to: CATV

Policy on Unauthorized Reception: N/A

Plans for Protection of Signal: N/A

Notes on Audio: N/A

Heartbeat Media Network
1700 Broadway
New York, NY 10019

Description of Services: video music entertainment and information programming for all television media covering rock, pop, rhythm and blues, country, and jazz. Individual three to five minute "lifestyle" segments will be available for distribution in the basic and pay cable markets as "continuity" programming.

Satellite & Transponder: Westar III

Operating Since: first programming projected to be on-the-air in early 1983.

Services Available to: programming will be sequentially distributed in broadcast, cable pay/STV, international and home video.

Policy on Unauthorized Reception: none described

Plans for Protection of Signal: none described

Notes on Audio: programming will be produced and made available in stereo.

HBO (Home Box Office)
Time-Life Building
Rockefeller Center
New York, NY 10020

Description of Services: movies, sports, and specials. Home Box Office is the first pay television service to go on satellite.

Satellite & Transponder: Satcom I, transponders 22, 24; Comstar D-2, transponder 9-H.

Operating Since: December, 1975

Services Available to: CATV

Policy on Unauthorized Reception: does not permit private Home Box Office reception of their signal. Home Box Office has been one of the industry leaders against unauthorized reception.

Plans for Protection of Signal: signal to be scrambled to prevent unauthorized reception.

Notes on Audio: N/A

Home Music Store
Digital Music Co.
1800 M Street NW
Washington, DC 20036

Description of Services: a multi-channel music service is offered.

Satellite & Transponder: Westar IV, transponder 22

Operating Since: spring, 1982

Services Available to: N/A

Policy on Unauthorized Reception: N/A

Plans for Protection of Signal: N/A

Notes on Audio: N/A

HTN (Home Theater Network)
465 Congress Street
Portland, ME 04101

Description of Services: family entertainment—G and PG movies.

Satellite & Transponder: Satcom I, transponder 21

Operating Since: September, 1978

Services Available to: CATV

Policy on Unauthorized Reception: N/A

Plans for Protection of Signal: N/A

Notes on Audio: N/A

Lifestyle
5200 South Harvard, Suite 215
Tulsa, OK 74135

Description of Services: background music is offered.

Satellite & Transponder: Satcom I, transponder 3

Operating Since: February, 1981

Services Available to: N/A

Policy on Unauthorized Reception: N/A

Plans for Protection of Signal: N/A

Notes on Audio: N/A

Modern Satellite Network
5000 Park Street North
St. Petersburg, FL 33709

Description of Services: public relations and advertising, motion pictures principally, in the consumer information area.

Satellite & Transponder: Satcom III-R transponder 22, 10 A.M. to 1 P.M. weekdays

Operating Since: January 29, 1979

Services Available to: advertisers only, private reception allowed.

Policy on Unauthorized Reception: the network doesn't care.

Plans for Protection of Signal: none

Notes on Audio: monaural only

The Movie Channel
1211 Avenue of the Americas
15th Floor
New York, NY 10036

Description of Services: all-movie pay TV service operated by Warner-Amex.

Satellite & Transponder: Satcom I, transponder 5

Operating Since: April, 1979

Services Available to: CATV

Policy on Unauthorized Reception: will not authorize viewing via home earth station.

Plans for Protection of Signal: N/A

Notes on Audio: N/A

MTV (Music Television)
1211 Avenue of the Americas
New York, NY 10036

Description of Services: concerts and music programming from Warner-Amex.

Satellite & Transponder: Satcom I, transponder 11

Operating Since: August, 1981

Services Available to: CATV

Policy on Unauthorized Reception: will not authorize viewing via home earth station.

Plans for Protection of Signal: N/A

Notes on Audio: Dolby FM stereo

The Nashville Network
2806 Opryland Drive
Nashville, TN 37214

Description of Services: it provides country variety entertainment.

Satellite & Transponder: Westar V

Operating Since: plans to begin operation in February, 1983.

Services Available to: CATV firms only

Policy on Unauthorized Reception: does not allow private reception of signals.

Plans for Protection of Signal: not stated

Notes on Audio: monaural only

National Christian Network
1150 West King Street
Cocoa, FL 32922

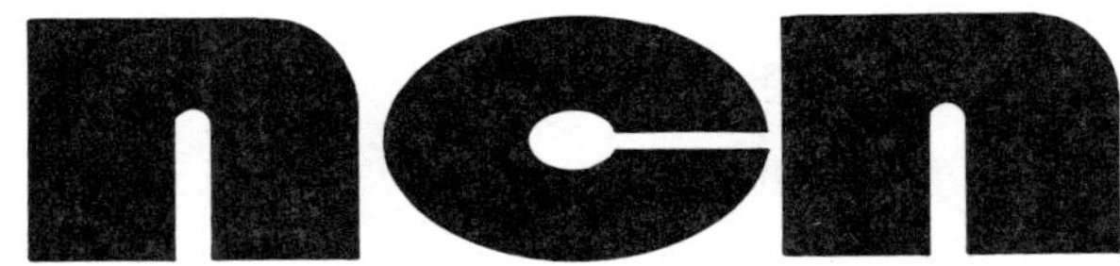

National Christian Network

Description of Services: Catholic, Protestant, and Jewish programming presented with open access to any religious group wishing to utilize the network's satellite distribution services.

Satellite & Transponder: Satcom IV, transponder 7

Operating Since: June 1, 1980

Services Available to: CATV, private citizens (no charge)

Policy on Unauthorized Reception: none

Plans for Protection of Signal: none

Notes on Audio: Monaural

National Jewish Television
2621 Palisades Avenue
Riverdale, NY 10463

Description of Services: it provides programming for the Jewish community.

Satellite & Transponder: Satcom I, transponder 16

Operating Since: May, 1981

Services Available to: N/A

Policy on Unauthorized Reception: N/A

Plans for Protection of Signal: N/A

Notes on Audio: N/A

Nickelodeon
1211 Avenue of the Americas
New York, NY 10036

Description of Services: Warner-Amex programming aimed at children and adolescents.

Satellite & Transponder: Satcom I, transponder 1

Operating Since: April, 1979

Services Available to: CATV

Policy on Unauthorized Reception: reception by home earth stations not authorized.

Plans for Protection of Signal: N/A

Notes on Audio: N/A

North American Newstime
1800 Peachtree Road
Atlanta, GA 30309

Description of Services: news service including the Women's Channel and television program guides.

Satellite & Transponder: Satcom I, transponder 6

Operating Since: May, 1981

Services Available to: N/A

Policy on Unauthorized Reception: N/A

Plans for Protection of Signal: N/A

Notes on Audio: N/A

North American/The Travel Channel
1375 Peachtree Street
Suite 320
Atlanta, GA 30309

Description of Services: twenty-four-hour travel and information with 10 minutes of news on the hour and half hour. Programming outlines best vacation spots in the world and gives travel tips on how to pack, exchange money, etc.

Satellite & Transponder: Satcom III, transponder 6

Operating Since: May 1, 1981

Services Available to: CATV, MATV, SMATV and LPTV. Private reception is allowed, "but it's very costly."

Policy on Unauthorized Reception: "if our signals are pirated and we can prove it, we will prosecute."

Plans for Protection of Signal: None

Notes on Audio: monaural

PTL (People That Love)
7224 Park Road
Charlotte , NC 28279

Description of Services: Christian-oriented programming featuring specials, news, and entertainment.

Satellite & Transponder: Satcom I, transponder 2

Operating Since: April, 1979

Services Available to: N/A

Policy on Unauthorized Reception: N/A

Plans for Protection of Signal: N/A

Notes on Audio: N/A

Penthouse Entertainment Television Network
TeleMine
888 Seventh Avenue
New York, NY 10106

Description of Services: films, features, and other entertainment with an adult slant.

Satellite & Transponder: Westar I, transponder 3

Operating Since: February, 1982

Services Available to: N/A

Policy on Unauthorized Reception: N/A

Plans for Protection of Signal: N/A

Notes on Audio: N/A

Private Screenings Inc.
330 West 42 Street
Penthouse Suite
New York, NY 10036

PRIVATE
SCREENINGS™ INC.

Description of Services: adult pay TV service consisting of erotic feature films and original entertainment specials.

Satellite & Transponder: previously on Westar III, now operating as a standalone.

Operating Since: May, 1980

Services Available to: CATV, STV, MDS, SMATV. It does not allow private reception.

Policy on Unauthorized Reception: none provided

Plans for Protection of Signal: none announced

Notes on Audio: monaural

Reuters Ltd.
2 Wall Street
New York, NY 10005

Description of Services: proprietary data signals

Satellite & Transponder: Satcom III-R, transponder 18

Operating Since: 1973

Services Available to: private TVROs, usual subscriptions from $99 per month.

Policy on Unauthorized Reception: "good luck to them."

Plans for Protection of Signal: none announced

Notes on Audio: no audio

Satellite Music Network
6500 River Chase Circle East
Atlanta, GA 30328

Description of Services: popular and country music are offered.

Satellite & Transponder: Satcom I, transponder 3

Operating Since: August, 1981

Services Available to: CATV

Policy on Unauthorized Reception: N/A

Plans for Protection of Signal: N/A

Notes on Audio: broadcasts in stereo

Satellite News Channel (SNC)
Group W Satellite Communications
41 Harbor Plaza Drive
P.O. Box 10210
Stanford, CT 06904

Description of Services: SNC is cable's only all-news, all-live, exclusive-to-cable news service. SNC offers continuously updated news summaries every 18 minutes. In addition, SNC utilizes five Westar transponders to deliver regional news in 24 regions across the country.

Satellite & Transponder: Westar V, five transponders

Operating Since: June 21, 1982

Services Available to: CATV only

Policy on Unauthorized Reception: to be announced

Plans for Protection of Signal: to be announced

Notes on Audio: monaural only

Satellite Program Network, Inc.
8252 South Harvard
Tulsa, OK 74136

Description of Services: "lifestyle" programming, including business, financial, health, art, exercise, nutrition, information/how-to, entertainment, international, classic movies, and variety.

Satellite & Transponder: Westar IV, transponder 11X. It will move to Spacenet in 1984.

Operating Since: January 22, 1979

Services Available to: CATV, MATV, SMATV, MDS, LPTV, etc. Service is free. "Per FCC rules and regs, we are not allowed to serve private uses."

Policy on Unauthorized Reception: did not respond

Plans for Protection of Signal: none announced

Notes on Audio: monaural only

Showtime Entertainment Inc.

Description of Services: not available

Satellite & Transponder: Satcom III-R transponder 10-West, transponder 12-East.

Operating Since: July 1976

Services Available to: CATV, hotels, motels, and SMATV in nonfranchised areas.

Policy on Unauthorized Reception: "Showtime considers unauthorized reception of its signal to be illegal. Yes, we prosecute."

Plans for Protection of Signal: Showtime's signal will be scrambled within 18 months of October, 1982.

Notes on Audio: monaural only

Showtime is a registered trademark of Showtime Entertainment and is used by permission.

SIN (Spanish International Network)
250 Park Avenue
New York, NY 10017

Description of Services: oldest Spanish language programming service via satellite. It features Spanish language series, specials, and news and does international transmissions under the name of Univision.

Satellite & Transponder: Westar III, transponder 8; Comstar D-2, transponder 4-H.

Operating Since: September, 1976

Services Available to: CATV

Policy on Unauthorized Reception: N/A

Plans for Protection of Signal: N/A

Notes on Audio: N/A

Spotlight
2951 28th Street #2000
Santa Monica, CA 90405

Description of Services: pay TV service for the Times-Mirror cable systems

Satellite & Transponder: Satcom I, transponder 19

Operating Since: May, 1978

Services Available to: CATV

Policy on Unauthorized Reception: N/A

Plans for Protection of Signal: N/A

Notes on Audio: N/A

The Weather Channel Inc.
2840 Mt. Wilkinson Parkway
Suite 200
Atlanta, GA 30339

Description of Services: a 24-hour-a-day live source of current national, regional, and local information presented live by on-camera meterologists and computer-generated graphics. Local information is made available through a satellite teletext transmission system using the vertical interval and the Weather STAR, an addressable amplifier-decoder.

Satellite & Transponder: Satcom III-R, transponder 21

Operating Since: May 2, 1982

Services Available to: the emphasis and priority is on cable television.

Policy on Unauthorized Reception: it is not directly stated.

Plans for Protection of Signal: the Weather Channel does not scramble its video or audio signal. Local teletext information can only be received through the use of the Weather STAR.

Notes on Audio: it does not broadcast in stereo.

Time Video Information Services, Inc.
Time & Life Building
Rockefeller Center
New York, NY 10020

Description of Services: testing a teletext service that consists of a wide variety of information and entertainment offerings such as news, sports, weather, games, and advertising.

Satellite & Transponder: undetermined

Operating Since: testing during 1983

Services Available to: not determined yet

Policy on Unauthorized Reception: firm owns property rights of electronic signals.

Plans for Protection of Signal: not announced/determined

Notes on Audio: N/A

Trinity Broadcasting Network, Inc.
P.O. Box A
Santa Ana, CA 92711

Description of Services: twenty-four hour-a-day Christian programming consisting of children's programs, Bible studies, musical variety shows, talk shows, etc.

Satellite & Transponder: Satcom IV, transponder 17

Operating Since: August 2, 1973

Services Available to: CATV, SMATV, etc. Service is not for sale, but it is offered free of charge. Reception by private citizens is allowed. An authorization form must be filled out.

Policy on Unauthorized Reception: "we offer our signal free of charge, and therefore would not prosecute anyone for receiving it."

Plans for Protection of Signal: we do not scramble.

Notes on Audio: monaural

Ultimedia Television (UTV)
22 08 Route 208
Fair Lawn, NJ 07410

Description of Services: consumer and general interest programming.

Satellite & Transponder: Satcom IV, transponder 23

Operating Since: spring, 1982

Services Available to: N/A

Policy on Unauthorized Reception: N/A

Plans for Protection of Signal: N/A

Notes on Audio: N/A

UPI Cablenews Wire
220 East 42nd Street
New York, NY 10017

Description of Services: text news stories from the wires of United Press International.

Satellite & Transponder: Satcom I, transponder 6

Operating Since: 1980

Services Available to: CATV

Policy on Unauthorized Reception: N/A

Plans for Protection of Signal: N/A

Notes on Audio: N/A

USA Network
208 Harristown Road
Glen Rock, NJ 07452

Description of Services: wide variety of sports, cultural, health and other kinds of programming. It includes baseball, hockey, and other sports, plus British series, children's programming, a fashion program, and health and physical fitness information.

Satellite & Transponder: Satcom I, transponder 9

Operating Since: September, 1980

Services Available to: CATV

Policy on Unauthorized Reception: N/A

Plans for Protection of Signal: N/A

Notes on Audio: N/A

UTV Cable Network
28-08 Route 208
Fair Lawn, NJ 07410

Description of Services: two-way interactive, broad-based television programming. Consumer and general interest programming with subscriber participation stressed.

Satellite & Transponder: Westar V, transponder 2X

Operating Since: start-up on January 10, 1983

Services Available to: CATV, SMATV. No private reception.

Policy on Unauthorized Reception: no policy established as of this writing.

Plans for Protection of Signal: none announced

Notes on Audio: monaural only

WFMT
5200 South Harvard, Suite 215
Tulsa, OK 74135

Description of Services: Chicago's arts-oriented radio station.

Satellite & Transponder: Satcom I, transponder 3

Operating Since: March, 1979

Services Available to: N/A

Policy on Unauthorized Reception: N/A

Plans for Protection of Signal: N/A

Notes on Audio: it broadcasts in stereo.

WGN
5200 South Harvard, Suite 215
Tulsa, OK 74135

Description of Services: movies, sports, news, and specials from Chicago's superstation.

Satellite & Transponder: Satcom I, transponder 3

Operating Since: November, 1978

Services Available to: N/A

Policy on Unauthorized Reception: N/A

Plans for Protection of Signal: N/A

Notes on Audio: N/A

WOR
3 Northern Concourse
Box 4872
Syracuse, NY 13321

Description of Services: New York's superstation carries syndicated shows, movies, and sports.

Satellite & Transponder: Satcom I, transponder 17

Operating Since: April, 1979

Services Available to: N/A

Policy on Unauthorized Reception: N/A

Plans for Protection of Signal: N/A

Notes on Audio: N/A

WTBS
1050 Techwood Drive NW
Atlanta, GA 30318

Description of Services: the original superstation. It carries movies, sports, syndicated programs, and news.

Satellite & Transponder: Satcom I, transponder 6

Operating Since: December, 1976

Services Available to: N/A

Policy on Unauthorized Reception: N/A

Plans for Protection of Signal: N/A

Notes on Audio: N/A

Chapter 5

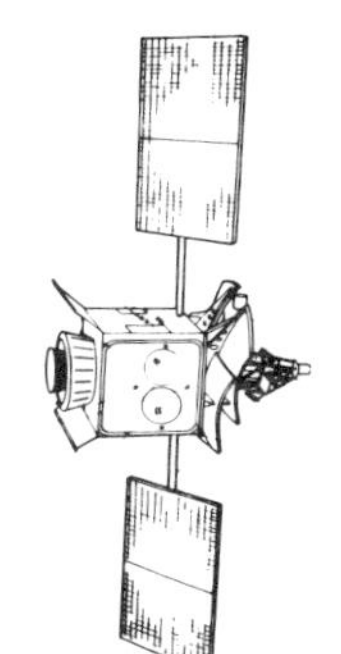

Getting Started in Satellite Television

Home satellite television is a pretty remarkable technology. Obviously, we're sold on it, and we think there are millions of people in this country who should be, too.

You were probably at least casually interested in the subject when you picked up this book. By now you may be anxious to put up your own home earth station. If so, our advice is to read this chapter carefully and then give the matter some thought.

Home satellite television today (late 1982) is where home computers were in 1976 or 1977. There are an increasing number of products on the market and some very rosy predictions for the future. The people who put up their earth stations in the next couple of years are going to be pioneers as much as the ones who built computers from kits six or seven years ago.

Home earth stations are not yet fully developed consumer goods like televisions or videotape recorders. The level of support for the user isn't all that it should be. The industry hasn't shaken out the weak products, and inferior products, and the overall ambience of the market is still more hobbyist-oriented than consumer-oriented.

If you decide to set up a home earth station you will need

a certain amount of pioneering spirit. You will probably encounter problems that will have to be overcome, and that there will be an element of challenge in the entire process.

The problems aren't insurmountable (at least not if you go about things right), and there's a real satisfaction in meeting those challenges. Ask anyone who built his own computer back in the mid-1970s. Putting in an earth station is not like buying a color television. You have to do more of the work yourself in the sense that you have more meaningful choices to make—choices that can make the difference between a pleasant experience and a smoothly functioning system and a nightmare with a system that works poorly or not at all.

We're telling you this not to frighten you, but because we want you to succeed at home satellite television. Your chances of success will go up if you understand that you have to be an active participant in the process. Given that, nearly anyone can succeed at installing a home earth station.

MONEY

The first thing most people wonder about satellite television is how much will it cost. The cost varies depending on several things.

You can set up a system right now for about $2,000 if you live in the right part of the country. Your picture quality will probably be poor, though, and your viewing choices will be limited. You can pay more than $20,000 for a system with more bells and whistles than with which you know what to do. Unless you have a lot of video experience or live in an area with real signal problems, you'll be paying for a lot more than you need.

A more realistic range would be $3,000 to $10,000 for a system with quality components and an antenna that can receive all the major American satellites. The amount depends on how much work you choose to do yourself and where you live.

Price is really a secondary consideration. First, you want a system that meets your needs and can give you a good picture. Beyond a certain point, trying to save money on your earth station gets you nothing but trouble.

Our advice is to find the minimum system that will give you the choice and picture quality you want before you start looking at price tags. If you can't afford that minimum system, forget about a home earth station for a couple of years. Equipment prices are dropping all the time as the market grows and new technologies are applied. Don't buy a system that's inadequate.

CAN YOU SET UP AN EARTH STATION?

In order for an earth station to operate, the antenna must have an unobstructed view of the satellite you wish to receive. If that isn't possible, you are stopped before you can even get started. If there's a mountain, a high-rise building, or even a big tree between you and the satellites, you won't get a good picture.

You may also need permission from your city's zoning department before you can put up the antenna. This is becoming more of an issue as cities include satellite dishes in their zoning ordinances.

You also need room for the antenna. This will be about as much area as a small swimming pool. If you want to receive more than one satellite, you must have room to swing the antenna.

Except for the pad to which the antenna attaches, the installation of a home earth station is not permanent. You can dismantle your earth station and take it with you when you move.

The area where you live will also have an influence. You can get good satellite reception in all 48 contiguous states, Alaska, and Hawaii (although there may be limitations there), but some parts of the country require better-quality equipment and a bigger antenna.

THE HARD TRUTH ABOUT SATELLITE TELEVISION COMPANIES

While researching this book, one of us had a very interesting conversation with a dealer. The man was reluctant to talk about the specifics of satellite television and seemed almost hostile. After some effort to build rapport, he loosened up a little and finally explained why.

"I was afraid you were fishing for information so you could go into competition with me," he said. He explained that at least one other person had approached him with many questions and then went out and opened a dealership.

To really appreciate this, you have to understand that the questions didn't concern the satellite television business at all. The questions were about owning a home earth station—questions that any intelligent consumer would ask.

Apparently this wasn't just paranoia on one dealer's part. A couple other dealers mentioned that the same thing had happened to them, or they knew someone to whom it had happened. One book on satellite television, written for complete novices, contains a chapter on how to become a home satellite television dealer.

We're all in favor of entrepreneurship, but you really have to wonder about the service you'd get from a dealer who had learned his business by asking a few casual questions or reading a book. Selling earth stations is not like selling refrigerators or used cars.

What is happening in the satellite television business is very similar to what happened when personal computers started to boom. Many companies started making them, and many people began selling them. These companies and people often didn't really know what they were doing.

There are good satellite television manufacturers and dealers. Some companies and dealers have standards as high as any you will find in any industry. There are also many dealers and a few manufacturers who aren't competent, and a few aren't even honest.

Because of a home earth station's nature, we advise very strongly that you get your system from someone who knows what he is doing. Don't become the guinea pig for someone who is learning the business or the sucker for a fast-buck operation.

There are a couple reasons why the field has these problems. One is that a home earth station is a big-ticket item. A dealer's cut on the sale of a single system can be several thousand dollars. To someone who isn't familiar with the economics of merchandising expensive consumer goods—particularly expensive high-tech consumer goods—it looks like a way to make a fast fortune. Sell just one system a week, and you're on easy street.

Very few dealers sell one system a week. They may only sell one a month or one every other month. There are also the expenses that go with the business. These include obvious things like store rent and less obvious things like customer support. The people who don't understand high-technology businesses or who are out for the fast buck will skimp on support, and their customers will suffer.

The home earth station business is still very new. The FCC didn't approve unlicensed receive-only earth stations until 1979. As a result, no one has much of a track record. The fact that a company has been in business less than a year doesn't mean as much in a brand-new industry as it would in a more established field. Because very few people know anything about the business, a dealer who wants to can get away with a lot more.

ASK AROUND

If you are seriously interested in home satellite television, your first step should be to ask around. Are there any successful installations in your area? By successful we mean an earth station that can receive all the channels for which it was designed. We also mean one whose owner is satisfied with the return on the time and money invested in the project.

Don't confuse a successful installation with a dish sitting in someone's backyard. There are many fewer successful systems than there are installed systems.

A previously successful earth station is your best guarantee that your earth station will be successful. Even an unsuccessful installation can teach you a lot. For example, a common cause of failure in setting up an earth station is using equipment that isn't up to the demands of local conditions. Perhaps you need a 12-foot antenna or a 14-foot one instead of an 8 or 10-foot dish. Maybe a 120-degree low-noise amplifier won't do in your area, and you will need a 90-degree one. Are there interference problems? Maybe a narrower feed horn or a notch filter will help.

If there are installations in your area, you should find out whom they were purchased from and what the owner thinks of the deal he got. It will help you to know who to deal with and who to avoid.

MAIL ORDER VERSUS A LOCAL DEALER

We have a strong preference for buying home earth stations from a local dealer—assuming that the local man is competent. There are many companies in the home satellite television business that sell entire earth station packages through the mail, and many others will sell you the components. Most are good companies with good products. You can often save a lot of money by purchasing your system through the mail. A mail-order company cannot give you the support that a local dealer can, particularly in installing the system.

There are still many areas that do not have a good home satellite dealer. Either there are no dealers at all, or you don't want to go to the available ones. Then you have to buy through the mail. Going to a dealer who won't provide service and support in your area doesn't do you any good.

Then there are the knowledgeable, self-reliant types who enjoy doing it themselves. If you're of that breed, you may want to dispense with a local dealer and save money by purchasing through the mail.

If you do purchase your equipment through the mail, get a phone number that you can call for help if problems should arise. Ideally, the manufacturer or mail-order dealer will have knowledgeable people on duty during business hours who can help you in case of trouble. Before you place an order, find out if that kind of support is going to be there.

DEALING WITH DEALERS

Find out if there are any dealers in your area. Look in the Yellow Pages under "Satellite Television."

Your "area" can be pretty large when shopping for an earth station. It's worth driving a hundred miles to look at equipment and talk to a dealer if he is willing to travel that far to support the equipment he sells. Many dealers are, particularly in the Midwest and West.

Leave your checkbook at home the first time you visit a dealer. First you need information, not an earth station.

Try to form an impression of how knowledgeable the dealer is about home satellite television. Does he seem to know what he is talking about, or does he just rattle on about all the wonderful shows you can receive? Can he answer technical questions about his equipment? Can he tell you what setup you need in your area?

Be prepared for some pretty amazing experiences. Don't be surprised if you know more about satellite television after reading this book than some of the dealers do. One salesman tried to switch from a horizontally polarized transponder to a vertically polarized one without rotating the antenna feed horn. The rotation control was sitting right next to the receiver, but he never touched it and could not understand why he couldn't get the signal. All the while he was talking about the "hundreds of channels" the system would bring in.

You may meet some extremely well-informed people who can tell you far more than you ever wanted to know about the insides of a satellite receiver. Many dealers got into the business through amateur radio or after being trained as engineers. It may be a little boring to listen to these people, but at least you know the expertise is there when you need it.

Find out what equipment the dealer carries. Some dealers only stock one brand of antenna, and some offer a choice. Pick up data sheets on the equipment if the dealer has any available.

Ask the dealer if he has any local customers to whom you can talk. He should be willing to give you some names if he thinks you are seriously interested in an earth station. Those people can tell you a lot about earth stations in your area and about your dealer.

Most importantly, find out what the dealer's warranty policy is on the equipment he sells and what sort of repair and adjustment facilities he has. If he can't support you, why bother?

Finally, contact the local Better Business Bureau and see if there are any complaints on file against the dealer. This piece of advice also applies when you are dealing mail order.

In the case of a mail-order dealer, it may cost you a long-distance phone call, but you will save yourself a lot of trouble.

Be wary of a dealer who pressures you to buy a system immediately, particularly by saying that it is on sale "for today only" or that he can get you a special price if you buy right now. Naturally the dealer is going to try to sell you a system. If he understands the business at all, he knows that a rational person is going to take the time to decide on a purchase this big.

RELIABILITY

Don't assume from our discussion about service and support that home earth stations are unreliable. On the contrary, they are quite reliable. Once they are working properly, they tend to keep working without trouble for a long time.

We stress support because it often takes a little fiddling to get home earth stations working well in the first place. A home earth station is essentially a component system. You have to play with component systems to get the best out of them. Because a home earth station has to handle signals in the gigahertz range, you can get many little problems.

To someone who understands earth stations and has some experience with them, it is not difficult to get one up and running—at least not usually. If you don't have that experience it can be a long, slow, frustrating process. You can't take your LNA down to the local television repairman and expect him to fix it.

LEGAL CONSIDERATIONS

Considering the amount of money involved in buying a home earth station, give some thought to the legal aspects of the transaction.

For openers, remember that the verbal statements of the dealer or his salesman are not binding. Any promises or representations not made in writing are almost impossible to enforce. You should have a written agreement with the seller that clearly spells things out.

If you are buying a complete system from a dealer who will also install it, you should try to get a written guarantee that it will perform properly. It is probably worthwhile to consult an attorney about the details of the purchase agreement. Your best protection is to buy your system from a reputable dealer who is willing to stand behind his product.

CHOOSING A SITE FOR YOUR HOME EARTH STATION

When we discuss a site for a home earth station, we are really considering a site for the antenna. This is an extremely important matter for good functioning of your system.

For most systems, antenna site selection is a matter of compromise. Many factors need to be considered, and in an ideal situation you could take all of them into account. If you live in a largely flat rural area, you may have an ideal situation. Otherwise you may have to make some compromises. This isn't necessarily fatal to your desire to have a home earth station, but it will probably influence some of your equipment choices.

The ideal antenna site would have a good view of the satellites that are above the horizon, be close to the house and on stable soil, would not be subject to excessive wind and snow loads, and would conform to any applicable zoning ordinances. Given a reasonably sized backyard, those conditions aren't hard to meet in most parts of the country.

Short of setting up an antenna and trying it, the best way to find out if your proposed site is suitable is to have a site survey done. There are companies that will do these surveys for a fee, but they tend to be rather expensive. A proper site survey involves soil tests, interference checks, and making sure the satellites are within the antenna's field of view.

Don't confuse a site survey with a line-of-sight check. This is simply seeing what satellites are visible to your antenna and where they are in the sky. You can make this check yourself.

Your dealer should be able to tell you how to do a line-of-sight check. He can advise you on problems with soil stability and wind and snow loads if they exist in your area.

CHECKING THE VIEW

A line-of-sight check is basically very simple. You locate each satellite that you're interested in and look at that spot. If there's nothing between you and the satellite, you're in business.

The trick is finding out where to look, but it's really not much of a trick at all. You can get the information from several different sources, including books in the public library.

Many manufacturers will tell you where to look. They supply this information with their receivers or antennas, and some will furnish you with the data before you buy the equipment.

Heathkit will sell you a satellite location package for $30 that explains the fundamentals of satellite television and shows you how to find the satellites in your sky. Write to Heath Company, Benton Harbor, MI 49022. Ask for kit SRA-8100-10.

There are also satellite locator services such as Satellite Computer Service (SCS) (1808 Pomona Drive, Las Cruces, NM 88001) that will send you a list of the locations of various satellites as seen from where you live. The service costs $19.95. Satellite Computer Service actually sends you a list of the apparent position of an object on the equator at each degree of longitude visible in your area. This doesn't guarantee that a satellite is there. If you want to find a newly launched bird, you don't have to write back to the company for a new printout.

You can also purchase computer programs that will calculate the information for you. For instance, the Satellite Center (Box 7104, San Francisco, CA 94120) sells a package of satellite programs for Apple and TRS-80 computers that includes a satellite locator.

You can find the appropriate formulas in astronomy and navigation books. A good one is *Practical Astronomy With Your Calculator* by Peter Duffett-Smith (Cambridge University Press). The chapter on coordinate systems gives you some useful background.

CELESTIAL NAVIGATION

In a sense the procedure for finding your satellites is the exact opposite of celestial navigation. The navigator uses heavenly bodies to find his position on the earth's surface. You need to use your position on the earth to find the heavenly bodies.

This is not nearly as complicated as it sounds. Basically, you have to find a couple of numbers and plug them into a pocket calculator that can handle trigonometric functions. You can buy a "scientific" or "slide rule" calculator for $30 to $50.

Because the satellites move in lockstep with the earth, they don't move in the sky. That eliminates about half the problem's complexity.

Satellite positions are given in degrees of longitude. As you probably know, position on the surface of the earth is given in terms of latitude and longitude. *Latitude* measures how far north or south you are, and *longitude* measures how far east or west. Latitude starts at the equator, which is 0. Longitude has its zero point at the Royal Astronomical Observatory in Greenwich, England. The reason Greenwich is the zero longitude line is the English invented the whole system at a time when they were masters of the seas.

Longitudes running west from Greenwich (toward us) are measured in degrees west longitude. Somewhere the other side of Hawaii, they meet the east longitudes at the 180-degree line.

Normally a position is given in latitude and longitude. Because geosynchronous satellites are right over the equator, their positions are given in longitude only.

If you had an earth station at the point where the Greenwich meridian crossed the equator, you could look up the satellite's position and point your antenna that many degrees west. If you were interested in a satellite at longitude 0, you'd aim the antenna straight up. To receive one at longitude 10 degrees, you'd aim your antenna 10 degrees due west, and so on. The only difficulty is that no one has an earth station at latitude 0, longitude 0, because that's off the coast of Africa.

You need to convert that longitude into something you can use.

When you don't know your position in latitude and longitude or it doesn't matter, you can express the location of an object in the sky in degrees of azimuth and elevation. *Elevation* refers to how many degrees above the horizon the object is. *Azimuth* is measured from 0 to 360 degrees with true north as zero. A navigator works by finding the position of an object in terms of azimuth and elevation and translating that into latitude and longitude. Let's do the job backwards. Assume you know your position in latitude and longitude and the position of what you're looking for in latitude and longitude. You need to translate that into an elevation and azimuth you can use to point the antenna.

To start figuring, you need to know the position of the satellite, which you get by looking in a table or in a program guide. Next you need to know your latitude and longitude. The easiest way to find your latitude and longitude is to ask. Try the public library, the nearest airport, or an amateur astronomer. You can also figure it closely enough from a state map if it happens to be marked in latitude and longitude.

The basic formula for finding the azimuth of a satellite is:

Azimuth = 180° + arctan (tan B/Sin A)
A = Latitude of earth station
B = Longitude of earth station—longitude of satellite.

If the difference between your longitude and the satellite's longitude is more than 82 degrees, you can quit. The satellite is over the horizon and useless to you.

If you're not sure how to use the trigonometric function keys on your calculator, look in the instruction book. Make sure you are working in degrees or that you convert the answer back into degrees when you go to use it. Some calculators figure their trig functions in radians. The ones that do have a key to convert degrees to radians.

There is one other point to remember about that formula. West longitudes are expressed as negative numbers. In other words, 135 degrees west longitude needs to be plugged into the formula as −135 degrees. Because we are in the Western Hemisphere, all longitudes will be negative numbers. (If you happen to live in the Southern Hemisphere, your latitude is a negative number, too.)

The number you got tells you how many degrees clockwise from true north you need to swing your antenna to point it at the satellite. Now you need to determine how high you need to point it to pick up the satellite. To do that, start by finding the *central angle* between your site and the satellite. The central angle is calculated like this:

$$\text{Central angle} = \text{arc cos} (\cos A \times \cos B)$$

You need the central angle for two reasons. First, you will need to use it to calculate the satellite's elevation. Second, if it is less than 82 degrees, you might as well quit here. The satellite is below the horizon.

Use the central angle to find something else called the *range* of the satellite. The range is how far away the satellite is, and it is figured like this:

$$\text{Range} = \sqrt{(R)^2 + (R+H)^2 - 2R\,(R+H)\cos C}$$
$$C = \text{central angle}$$

R is the radius of the earth in miles (3,957). H is the distance from the earth to the satellite in miles (22,245).

Now that you know the range and the central angle, you can calculate the elevation, which was what you wanted to know in the first place. It is calculated like this:

$$\text{Elevation} = \text{arc cos}\,[(RA^2 + R^2 - (R\text{-}H)^2)/2 \times RA \times R] - 90^\circ$$
$$RA = \text{range}$$

You now know the elevation and azimuth of the satellite.

LOOKING FOR YOUR SATELLITES

You have a list of the satellites you are interested in and

the azimuth and elevation of each one from your location. You need to find out if you can see them from where you want to put your antenna.

If you have access to surveying instruments and know how to use them, this is no problem. If you don't, it's simple enough to cobble something together that will do the job. All you need is something to sight through and a way of setting it to a particular azimuth and elevation. You can make something with a couple of cheap protractors, the cardboard tube from a roll of paper towels, some string, a couple of weights, and a photographer's tripod. In a pinch you can dispense with the tripod and use a stepladder.

The other thing you need is a compass, preferably the kind with a rotating bezel marked off in degrees. In order to find the azimuth, you need to know true north. True north differs from magnetic north depending on where you are. You can find the amount of correction you need to make by consulting a book for a map, asking at the airport, or checking with the public library.

To find true north, point the compass needle north and offset the bezel by the amount of deviation. The north mark on the bezel will then point to true north.

To turn the collection of odds and ends we mentioned into a satellite sighter, you need to tape the cardboard tube to the head of the tripod in such a way that the tube can swivel and rotate with it. A protractor taped to the side of the tripod head can show elevation with a weighted string acting as a plumb bob/pointer. A 360-degree (full-circle) protractor can be attached to the tripod head to act as an azimuth indicator.

Finding a satellite with this gadget is simple. Set it so the tube is pointing at true north and rotate the tripod head until you get the proper azimuth reading off the 360-degree protractor. Then elevate the head until the string crosses the appropriate degree mark of the elevation indicator.

Look through the tube. If all you see is sky, you're in great shape. If there's a building or a tree in your field of view, you have a problem. You won't be able to receive the satellite's signal from that location.

Repeat the process for every satellite in which you're interested. You might try it from different points to find the one where you can "see" the most satellites.

COPING WITH VISIBILITY PROBLEMS

The further north you go, the more likely you are to have line-of-sight problems. The reason is that the satellites are lower in the sky at higher latitudes. That means the antenna is pointed closer to the horizon and is more likely to encounter obstructions.

The easiest thing to do about a visibility problem is to move the antenna location. This may not always be possible, or it may be too expensive. In that case you can consider removing the obstructions.

The third thing you can do is just live with a limited field of vision. If most of the programming you're interested in is on one or two satellites, it may not be worth the time and money to do anything if other satellites are blocked.

CHECKING FOR INTERFERENCE

Like the satellites themselves, you can't see microwave interference. Unlike the satellites, you don't know if it's there until you check.

The two most common sources of interference at microwave frequencies are television transmitters and communications links. The location of those sources is usually pretty easy to determine.

Are there any television stations near where you live? If so, they might prevent you from getting a quality picture, especially if they lie between you and the satellite you want to receive. You can find out the location of television stations by looking in the phone book.

Most microwave communications nets in this country are handled by the telephone company. You can call the telephone company and ask if you are between any of their relay stations. Be prepared to be patient and to tell your story several times, though. It's not the kind of request they get every day.

The other way to find out about interference sources is to ask your dealer or others with earth stations. Remember that the interference may be highly localized. Unlike radio-frequency interference, microwave interference tends to be geographically confined.

The surefire way of discovering potential interference is to hire someone to come out and test with the appropriate equipment. Meters for measuring microwave interference aren't that common, and they can be expensive. Some satellite television dealers can perform such measurements.

You are generally not going to have interference problems in a rural area unless you live in the path of a communications relay station. Urban and suburban areas are more problematical because of the larger number of sources.

If you live in an area with potential interference problems, you may need a better LNA and a feed horn designed to minimize interference.

SIGNAL STRENGTH AND ITS EFFECTS

Because satellites transmit directionally, some parts of the country are definitely better than others for receiving signals. Most satellite antennas are aimed to put their strongest signals into midwestern America. The further you get away from there, the weaker the signal becomes.

For the purposes of satellite television, the ideal place to live in this country is about 200 miles south and west of Chicago. The worst are south Florida, south Texas, and (for a different reason) northern Maine.

This is relative. None of these "fringe" areas are really bad for satellite television in the way that a fringe area is bad for normal television. At the most it means you need a bigger antenna and a better LNA.

Not all satellite transmitters are focused in this pattern. Some will have different signal strengths in the same area because their beams are focused differently. Some of the Canadian Anik satellites, for instance, are designed to beam signals to the far north. Other satellites may have their beams set to concentrate on the east or west coast instead of aiming at the entire nation.

The problem that northern states such as Maine have is for them the satellites are low on the horizon. We have already mentioned that this increases the probability of obstructions, but there is another difficulty, too. The amount of ground noise the system picks up increases as the elevation of the antenna decreases.

FOUNDATIONS AND WEATHER

Wherever you put your antenna, it must be firmly anchored. It cannot be overstressed by wind and snow. An earth station antenna is big and seems heavy when you're putting it in, but it has plenty of surface area to catch wind and collect snow.

To function properly, the antenna must be properly aimed. If the base shifts, the antenna's aim will be thrown off. This is particularly a problem with a polar antenna mount, because you can't compensate for it with the antenna rotation controls. You need a good, solid base on which to put your dish.

That means a concrete pad or a series of concrete piers extending down below the frost line. This doesn't have to be massive, but it had better be sturdy or you are going to have trouble. Your antenna's manufacturer should include instructions on preparing the footings with the antenna, and the dealer should know what it will take.

Occasionally you will come across an antenna mounted on someone's roof. This is definitely not a do-it-yourself project. It takes an architect to do the necessary stress calculations and to come up with a structural design that will keep your antenna from coming downstairs to join you at the dinner table, or taking off across country in a high wind. Some lightweight antennas can be mounted on a roof, but they usually require a special structure to protect them from the wind.

Properly anchored, your antenna isn't going anywhere. Most antennas are designed to stand up to 100-mile-an-hour winds when they are correctly supported.

If you live in a windy area, consider giving your antenna some protection from the wind. Putting it on the lee side of a house or other building is a good idea, if possible. Another good plan is to provide some type of windbreak. This can be a planting of tall trees and shrubs or another kind of structure. Another possibility is to put the antenna in a pit with sides sloped so they don't interfere with the line of sight to the satellites. If you use a pit, give some serious thought to drainage.

Snow is probably less destructive but potentially more troublesome. If you live in an area with hard winters, think of your satellite dish as a 10-foot snow scoop. The antenna isn't exactly delicate, but the snow does need to be cleaned off it. Make sure that the antenna rotating mechanism and the feed horn rotation (if that is how you separate the polarized signals) are protected well enough to continue to work in cold.

In areas where wind and snow are problems, you might consider an antenna made of metal mesh. These antennas provide good reception and aren't as susceptible to wind and snow damage.

SELECTING THE RIGHT ANTENNA

There are two basic kinds of antennas used for home satellite systems today—parabolic and spherical. A *parabolic* antenna has a parabolic surface to focus the signal at a single point. The shape produces the maximum amount of signal concentration for a given size antenna, and it is the most common shape of antenna.

A *spherical* antenna produces less gain, but it has a very interesting geometric property which makes up for that in some situations. Recall from basic geometry that a parabola has only one focus. A signal coming from any direction that impinges on the parabola will be reflected to the focus. A sphere will focus reflections in different places depending on the angle of arrival of the impinging signal.

Astronomers have understood the difference for a long time, and they use it in designing telescopes. Most optical telescopes use parabolic mirrors and focus the light on a fixed mirror to reflect it to the eyepiece. You aim the telescope by pointing the mirror at the object you want to observe. Some very large radio telescopes, like the one at Arecibo in Puerto Rico, are too big to be swung. Instead the reflecting surface remains stationary, and you aim the telescope by moving the receiver.

You can do exactly the same thing with a spherical antenna. Generally the antenna has only a limited amount of mobility, but the feed horn and LNA can be moved to pick up different satellites.

A reasonably good system of this sort can pick up signals off six or seven adjacent satellites. The ones that are off-center will not be as strong as the one the antenna is pointed at directly, but they will still be adequate in most parts of the country.

Spherical antennas are becoming increasingly popular for home earth stations, particularly because a couple of companies sell them as kits. Because spherical antennas do not have to move on their bases, they can be made lighter than parabolic dishes. This also makes them more suitable for roof mountings.

One problem with a spherical antenna is that it is harder to automate satellite switching. Most parabolic antennas feature an optional motor drive that will move the dish from one satellite to another. This is harder to do with a spherical antenna because the feed horn/LNA assembly will typically be 15 feet away from the antenna surface, and it must move through a much wider arc.

Once you decide on the kind of antenna you want, the next question is the size. Parabolic antennas usually come in 3, 4, and 5-meter sizes. Three-meter dishes are the most common for home stations. The 4-meter size is more commonly used in areas where signals are weak. The 5-meter size is really a commercial product that is sometimes used in a real problem area.

Bigger is generally better when it comes to satellite antennas, but a bigger antenna is also more expensive and means more problems in erecting it. Prices will vary from about $1,000 for a 3-meter antenna to between $5,000 and $10,000 for a 5-meter dish. The smaller sizes can usually be put up by a crew of three or four men on a prepared base with little or no special equipment. The bigger antennas may require the use of a crane.

Spherical antennas usually come in 8, 10, and 12-foot sizes, although some, such as Downlink's Skyview, are modular. Several antennas can be fastened together to produce a much larger antenna. Downlink says that you can assemble a 20-foot dish out of their antennas.

The antenna may be made of metal, fiberglass, or metal mesh. All have proven satisfactory. Whatever material the antenna is made of, its surface must be to fairly tight tolerances. This is especially true of parabolic antennas.

There are two basic kinds of antenna mounts. The cheapest and easiest to understand is the elevation-azimuth (EL-AZ) mount. This kind of mount has one adjustment for elevation and another for azimuth. Its biggest advantage is that it's cheap. A secondary advantage is that it's more flexible, so it is easier to allow for misalignment of the antenna.

A polar mount is tilted so it only needs one adjustment to move from one satellite to another. The angle of the tilt is determined by the latitude of the earth station and is established when the antenna is installed.

The polar mount's biggest advantage is that it is easier to move, particularly if a remote control antenna rotator is used. It is more expensive and harder to make adjustments if things get out of alignment due to foundation settling or other causes.

THE LOW-NOISE AMPLIFIER AND FEED ASSEMBLY

The *feed assembly*—that piece of ductwork out at the focal point of your antenna—has a major bearing on the performance of your entire earth station (Fig. 5-1). Physically the feed assembly doesn't look like much, but its size and shape has to match your system and performance requirements to give you the best possible response.

The design of the feed horn represents a trade-off between sensitivity and interference rejection. Generally the wider the angle of the feed horn, the more sensitive it will be and the more interference it will pick up.

The one thing a feed horn must do is match the LNA to which it is coupled. A mismatch here can cause a serious loss of signal strength.

On most home earth stations the polarity of the signal received is controlled by rotating the feed horn and LNA. This can be done manually or by remote control. Many remote controllers are adapted from television antenna rotators. A good rotator needs to be smooth in operation and reasonably precise. A rotator that hangs up halfway will produce interference problems. The rotator should also be adequately protected from the elements, so it will not be foiled by snow and ice.

The low-noise amplifier is one of the most important and expensive parts of any earth station. If you look at the schematic of an LNA, this seems odd. The circuit is fairly simple, and there aren't that many parts. The reason for the high cost is the exacting tolerances of the component, required for the microwave frequencies.

A LNA is a small metal box with the feed horn bolted to one end and a coaxial cable coming out of the other. It is a solid-state device and usually quite rugged.

LNAs are rated in degrees Kelvin. The lower the number, the better the LNA. The standard for home earth stations is a 120-degree LNA, although 90-degree units are used in areas with weak signals.

Noise temperature is the most discussed number when you're talking about LNAs, but it isn't the only important one. There are several models of 120-degree LNAs on the market, but they are not all equal. Another number you need to look at is the gain.

Gain on an LNA works just like the gain on a stereo amplifier. The more gain you have, the more the signal can be amplified. Gain is measured in decibels (dB). Because the decibel scale is nonlinear (actually it's logarithmic), even a small change in the gain in dB can indicate a large change in performance. An increase of 3 dB means that the output strength doubles.

The difference in gain in commercial LNAs with the same noise temperature can be a lot more than just 3 dB. In

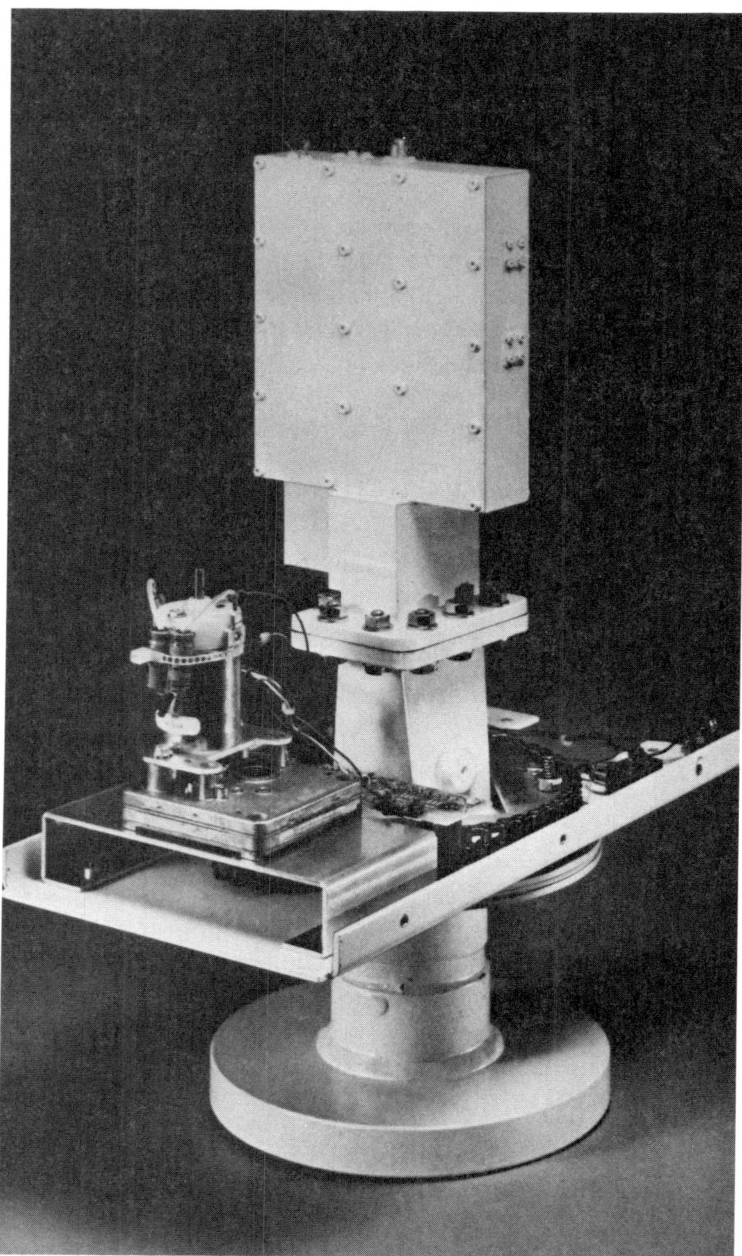

Fig. 5-1. The low-noise amplifier, down converter, and rotatable feed assembly from a Heathkit home earth station (courtesy Heathkit).

some cases it is 10 dB or even more. This can make a significant difference in the performance of a home earth station.

Most manufacturers include basic technical specifications in their literature. A salesman might not know offhand, but the information will be on the fliers he hands out.

CABLES, CONNECTIONS, AND CONVERTERS

Coming out of the LNA's back is one of the most troublesome parts of a home earth station. Special coaxial cable connects the LNA to the converter.

The signal from the LNA is still at 4 GHz. It is a lot stronger than what went in through the feed horn, but it still needs very careful handling. Signals at those frequencies aren't easily handled by a cable. They tend to leak off, attenuate, and do other strange things. That makes this section of cable extremely important for the successful functioning of a home earth station.

We emphasize this because most people aren't familiar with this type of equipment. In most pieces of electronic equipment a wire is a wire. If both ends are properly connected and the wire is correctly insulated, it will work.

At 4 GHz it is something to look at carefully. You need to use the right kind of cable, keep it as short as possible, and make sure that it is properly terminated or connected. Otherwise you may end up with some problems that will drive you nuts while you try to track them down.

The cable used to connect the LNA to the converter is special stuff. Because it is expensive, some people have been tempted to substitute ordinary coaxial cable. Ordinary coax doesn't work as well and can seriously degrade the performance of your system.

Make sure the cable connections are properly made. This is a simple process, but it must be done right or you can have a serious signal loss at the connections. If you are installing the system yourself, either use ready-made cable with the connectors on the ends or practice making connections until you have mastered the technique.

Some manufacturers offer low-noise converters (LNCs)

that combine the LNA with the converter stage on the antenna. In some cases the converter is in the same box as the LNA at the focus of the antenna. Other designs put the converter in a separate box on the antenna frame and connect the two with a short piece of cable.

Low-noise converters are almost mandatory where there is going to be a long run from the antenna to the house. The signal that comes out of the converter is around 70 MHz, and it is much easier to handle on cable.

RECEIVERS AND CONVERTERS

If you don't have a LNC, you will need to get a separate converter. This device takes the 4-GHz signal and reduces it to a frequency your satellite receiver can handle.

Some converters are designed to be mounted outdoors on or near the antenna. This keeps down the length the 4-GHz signal has to travel before conversion. Others are designed to be mounted next to the satellite receiver. Some receivers even have built-in converters.

The receiver is the part of your home earth station you will interact with the most. A satellite receiver looks like a piece of stereo equipment. Many receivers are mounted in wood or brushed aluminum cases, so they blend in with your stereo components.

The important thing about a receiver is what is inside the case. There can be several things in there depending on the receiver. The case can include a converter, a demodulator, and filters. It must include equipment for selecting channels.

The current trend in satellite receivers seems to be to build them in two parts. One part, incorporating the down converter, is mounted on the antenna. Sometimes this section does the channel selection as well. The other part sits in your living room and includes the controls. The two parts are joined by a coaxial cable. The cable should be of high quality, but it doesn't have to be to the specifications of the cable that connects the LNA to the converter.

The satellite receiver is probably the piece of the system that most reflects the consumer electronics philosophy.

There are many features available on the receivers, and they are definitely styled like consumer products.

There is one feature we recommend when buying a home earth station, whether you get it as part of the receiver or not. That is an alarm to tell you if someone is tampering with your antenna or LNA.

A home earth station has a lot of expensive electronics sitting out in the open. Vandals have been a bigger problem than thieves for earth station owners, but security is advisable.

Several satellite receivers incorporate alarm circuits. If the one you select doesn't, consider buying an alarm system separately. If you're not worried about alarming the neighbors, you can also hang high voltage warning signs all over the equipment. The electricity available at a typical earth station antenna wouldn't light a flashlight, but people don't know that yet.

The critical things for a satellite receiver are its ability to select the proper signal and send a clear, sharp picture to your television set. One good way to check out a satellite receiver is to see that model in operation in your area.

BUILDING A HOME EARTH STATION YOURSELF

We have said that the components of a home earth station are simple, and we've also indicated that the finished equipment is expensive. The obvious question, then, is can you save money by building the equipment yourself. You can if you have the skill and knowledge.

Home satellite equipment is emphatically not a beginner's project. In our opinion a person lacking the necessary training and test equipment would be foolish to attempt to build an earth station from scratch or by modifying surplus parts.

If you are a reasonably skilled kit builder, there are many components you can build yourself. The most obvious is the satellite receiver. There are receiver kits on the market from Heath and KLM Electronics Inc. (Box 816, Morgan Hill, CA 95037) as well as other sources. The May and June 1982 issues of *Radio-Electronics* magazine carried an article on a do-it-yourself receiver kit.

Anyone thinking about ordering a kit should look at those issues of *Radio-Electronics* to get an idea of what is involved. Among the test equipment needed to align the receiver are a CATV-type sweep generator with markers at 60, 70, and 80 MHz and an extended-response oscilloscope with an external-trigger input. The article also recommends using a calibrated spectrum analyzer to help with alignment.

If you're comfortable using that kind of gear, then the do-it-yourself route may be for you. If not, you might be better off buying components or a turnkey system.

Some other kits are simpler because they come with the high-frequency components already aligned. None of them is a beginner's project.

LNAs are the same way. Gillaspie and Associates (177 Webster St., Suite A455, Monterey, CA 93949) sells a LNA kit for less than $50. The kit does not include the special transistors required and some other components. They have to be purchased separately, bringing the total price up to between $150 and $200. There aren't many parts to the kit, but careful work is required to assemble it properly.

The finished product has a minimum gain of about 20 dB, compared to 50 dB or so for most ready-made LNAs. This is low, but it is adequate in some areas. The company says that the circuit's performance can be improved by knowledgeable tweaking.

There are a couple of spherical antenna kits on the market that appear to be fairly easy to construct. Both of them use a frame made of angle iron, wood strips to establish the shape of the dish, and metal mesh as the reflecting surface. The wooden strips are bolted to the metal frame. The curvature of the antenna is set by carefully tightening the bolts until the wood is bowed to the proper shape. If you want a spherical antenna, these kits are definitely worth considering.

For many people, building an earth station from a kit is attractive because it will save them money. We don't find that to be a convincing reason to try to build it yourself.

We have both built kits in our time. While we've enjoyed

the experience, we are somewhat skeptical of the economic benefits. A kit seller is in effect offering to pay the buyer for doing final assembly on the product. You can expect the difference between the kit price and the assembled price to be an indication of what it would cost to do the assembly at the factory with special equipment and skilled help. If you buy a kit solely to save money, you are essentially betting that you can beat the seller at his own game. In practice that usually means that you are valuing your time at little or nothing. If you are building equipment from kits because you enjoy the experience or because you want to really understand the finished product, that is a different matter. You're probably better off economically buying your gear ready-made.

TYPES OF INTERFERENCE

More than 50 percent of the home earth stations suffer some loss of picture quality on at least one channel. About 20 percent of them are totally inoperable because of microwave interference.

You don't have to simply accept interference. There are many things you can do to alleviate the problem.

There are two main kinds of interference you may encounter. One is *microwave interference* from other signals in the 4-GHz range. The most common cause is telephone microwave transmissions and Air Force radar. Occasionally you'll get some interference from a radio amateur operating in a nearby band. The other kind is *intermediate frequency interference* in the vhf (broadcast television) band.

The second kind of interference usually shows up when you are near a strong television station. It shows up as the station's television picture on your screen when you're operating your earth station. This is the easiest kind of interference to diagnose and potentially the easiest to cure. The signal is not coming in through your antenna. It is sneaking in through the receiver or the coaxial cable that connects it to the LNA or converter. Make sure all your connections are good and that everything is properly grounded. You might enclose the receiver case in metal mesh or install a power line filter.

Microwave interference shows up in one of two forms. Snow on the screen—so-called "sparklies"—is usually caused by a telephone microwave transmitter. These transmitters operate on frequencies inside the satellite television band and may come and go depending on the amount of telephone traffic. The other form of microwave interference is from an out-of-band transmitter such as a radar set. It acts to increase the noise-to-signal ratio and results in a smeared, cloudy, degraded picture—or no picture at all if the source is strong enough.

The first solution in all cases of microwave interference is filters. The out-of-band interference can usually be handled by putting a microwave filter between the LNA and the down converter. A properly designed filter will pass the frequencies you are interested in and cut transmission sharply on out-of-band signals. These filters are available from dealers and companies that sell earth station equipment.

Telephone interference won't be affected by a microwave filter, because that noise is in the band in which you're interested. The only signal you want is the one that comes out of your converter at around 70 MHz. If you put filters on the intermediate frequency stage of your earth station at 60 and 80 MHz, you will keep Ma Bell from reaching out and touching your television picture. The kind of filters you need are called traps or *notch filters*—so-called because they only filter a narrow band of frequencies. You need filters that have at least a 30-dB notch depth.

These filters are easiest to install on a system that has the receiver in two parts or a down converter separate from the receiver. You can simply attach them to the cable between the two parts. If you have an all-in-one receiver/converter, you will have to get someone to open the case and tap into the intermediate frequency stages to install the filters.

If you cannot solve your interference problems with filters, you may have to upgrade your equipment. Perhaps a narrower feed horn will eliminate unwanted signals. You may need a more sensitive receiver, a bigger dish, or a better LNA. This can get expensive. Talk to your dealer about what you need. If you are having serious problems with interference, the chances are excellent that other installations in your area are experiencing them too.

FINAL THOUGHTS

After we reread this chapter, we seriously considered rewriting it. It seemed to us that we had put too much stress on the negative aspects of setting up a home earth station.

We decided not to rewrite, but we should point out that in spite of all the problems and pitfalls we have discussed in this chapter, setting up a successful home earth station is not really difficult. We have spent plenty of time discussing things that can go wrong because forewarned is forearmed. If you know about these things, you can avoid them.

There are earth stations that don't work very well. We would even expect that the installation process will not go completely smoothly. If you don't get discouraged and give up, and if you get the right kind of help when you need it, you can have a successful installation.

Remember that once the initial adjustment period is past, your system should keep working for years. Occasionally you'll need to do things like grease the bearings of the antenna. Once home earth stations are up and running, they are extremely reliable.

Chapter 6

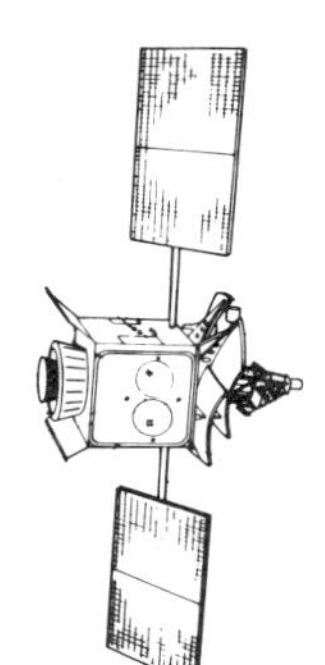

Products

Home satellite television is in a state of flux. This is particularly true of the equipment on the market. It seems like every month new companies come into the marketplace offering LNAs, antennas, receivers, and other gear.

This makes it difficult to write about what is available. By the time you read this book, a product may have been superseded or the specifications may have changed radically. It is difficult to talk meaningfully in general terms. The industry changes so quickly it is hard to sort out the philosophies or even the general approaches toward satellite television equipment. There are a few things we can say, however, based on the industry as it appears in late 1982.

The first and most obvious is that it is an industry and not just a collection of amateurs selling to other amateurs. The products are becoming more professional and so is the approach to marketing them. There is a growing concern for what the viewer wants and an increasing attempt to make life easy for him. The days when you had to go outside in the snow to shift the antenna or change the polarity of the LNA are gone.

The second observation is almost as obvious. The price of home earth stations has dropped significantly in the last three years, and it is going to continue dropping. The first home earth stations were in the $10,000 to $20,000 range and required the skill of a video engineer to operate. A modern earth station sells for $3,000 to $10,000 and can easily be operated by a 12-year-old.

We predict there will be substantial price decreases over the next three to five years, although the rate of decrease will be slow. We believe that within five years someone will bring out a completely assembled earth station for less than $1,000. It won't look like today's earth stations, but it will pull in all the available satellites just fine.

We also predict (although with less certainty) that the $1,000 earth station will feature a lightweight antenna made by stretching metallized plastic film umbrellalike over a rigid frame, a LNA/converter made with *gallium arsenide* (GaAs), integrated circuits, and a receiver with built-in filters and modulator. Most of the circuitry will be in three or four-integrated circuit packages. There will probably be a microprocessor built into the package to provide active noise rejection and possibly computer image enhancement.

The basis for this prediction is the work being done on GaAs integrated circuits right now. As these are developed and the market for satellite television equipment grows, the prices of equipment will tumble further.

The price of a home earth station may drop even further. There is no inherent reason why the curve couldn't continue down as steeply as it has in the last three years. You could then possibly buy an earth station in 1986 or 1987 for $300.

Spherical antennas seem to be gaining in popularity. The antennas are fairly new on the market, but they are attracting a great deal of interest, particularly in areas with high signal strength and low interference.

Another item that is growing in popularity is the LNC. The idea of putting the converter as close to the LNA as possible is gaining more recognition. Whether the unit is sold as a LNA converter or as an LNC, there is a trend toward mounting the converter package on the antenna frame. The all-in-one LNC, where the converter is in the same package as the LNA circuitry, is becoming more popular. There has to be some provision for mounting microwave filters in between the amplifier and converter stages for installations that have a problem with microwave interference.

Do-it-yourself kits are appearing on the market in greater numbers. The new ones tend to have the difficult parts of the job (such as aligning the microwave stages) already done. This makes them much more practical for kit builders.

There also seems to be a greater tendency for companies to build equipment for the home earth station market instead of just selling equipment designed for commercial operations. In many cases the same companies are making the product, but they are recognizing the home market as a separate market segment.

Following is some of the home earth station equipment available in the fall of 1982 (Fig. 6-1). Remember that this is a rapidly changing field. There is no assurance that the products, specifications, and (especially) prices will hold by the time you read this. We have tried to make the information here as accurate and complete as possible, but the way this business changes there is no guarantee of either accuracy or completeness. If you have any questions about a product or if you are thinking of purchasing, contact the manufacturer, his representative, or your local dealer for up-to-date information.

A-B Electronics
1782 West 32nd Place
Hialeah, FL 33012
(305) 887-3203
Parabolic antennas.

ADM (Antenna Development and Manufacturing)
Box 1178
Poplar Bluff, MO 63901
(314) 785-5988 or (314) 686-1484
Complete line of antennas including an 11-foot model that can be expanded to 13 feet and a 20-foot model. Antennas are of aluminum and steel construction. They feature polar mount

Fig. 6-1. The Heathkit satellite receiver is designed to fit into a home in the same way a stereo set does (courtesy Heathkit).

with optional remote control drive. Remote control rotating feed is standard.

American Micro Supply
500 South 9th Street
Cambridge, OH 43725
(614) 439-1552
Complete earth stations.

American Microwave Technology
Box 824
Fairfield, IA 52556
(515) 472-3174
Low-noise amplifiers from 70 to 110 K. Extremely high-quality equipment mostly for cable operators.

Amplica Inc.
950 Lawrence Drive
Newbury Park, CA 91320
(805) 498-9671
Low-noise amplifiers and 4-GHz line amplifiers. One of the major manufacturers of LNAs.

Andrew Corp.
10500 West 153rd Street
Orlando Park, IL 60462
(312) 349-3300
Antennas including 3-meter, 8-meter, and 10-meter models.

Anixter Brothers
4711 Golf Road
Skokie, IL 60076
(312) 298-9420
Complete home earth stations including antennas, LNAs, and receivers. Trailer-mounted models are available. Station with 3-meter dish listed for $4,000 to $7,000. A 5-meter dish and upgraded equipment station are priced at $7,600 to $12,000.

ATV Research
13th and Broadway
Dakota City, NE 68731
(402) 987-3771
Parabolic antennas.

Automation Techniques, Inc.
1846 North 106th East Avenue
Tulsa, OK 74116
(918) 836-2584
Satellite receivers. GLR 550 features two audio channels for direct or matrix stereo, push-button transponder selection, rf signal meter, and a weatherized tuning module with LNA power block.

Avantek Inc.
3175 Bowers Avenue
Santa Clara, CA 95051
(408) 496-6710
Products include Simulchannel Earth Station video receiving system, LNA/down converters, microwave semiconductors, and others.

AVCOM of Virginia Inc.
500 Research Road
Richmond, VA 23236
(804) 794-2500
Satellite receivers and complete earth stations. The Com-3 and Com-3R receivers with the Com-3R remote unit offer 24-position switch-selectable tuning, and switchable video polarity. AVCOM also carries accessories, LNAs, and complete systems.

Birdview Satellite Communications Inc.
Box 963
Chaunte, KS 66720
(316) 431-0400
Complete home earth stations.

Cayson Electronics
Route 3, Box 160
Fulton, MN 38843
(601) 862-2132
Complete home earth stations and components.

Central Satellite
Box 684
Jasper, IN 47546
Complete home earth stations.

Channel Master
Ellenville, NY 12428
(914) 647-5000
Complete home earth stations. Four models ranging from $5,990 to $7,590.

Channel One Inc.
Willarch Road
Lincoln, MA 01773
(617) 259-0333
Complete home earth stations.

Chaparral Communications
Box 832
Los Altos, CA 94022
(415) 941-1555
Feed horns for spherical and parabolic antennas.

Robert M. Coleman
Route 3, Box 58-A
Travelers Rest, SC 29690
Starview satellite television receiver. Also produces circuit boards for building earth station equipment.

Communications Plus
3680 Cote Vertu
The Bazaar Center
St. Lawrence, PQ Canada, HR41P8
(514) 337-7255
Satellite receiver and antennas.

Comtech Antenna Corp.
895 Central Florida Parkway
Orlando, FL 32769
(305) 892-6111
Satellite receiver, antennas, and feed horns. The model 650 receiver has remote control capability, 24-channel LED display, and one-year warranty.

Comtech Data Corp.
350 North Hayden Road
Scottsdale, AZ
(602) 968-2433

Complete line of home earth station equipment including receivers and antennas.

Cresscom Inc.
10-B Washington Avenue
Fairfield, NJ 07006
(201) 575-4184
Cable and connectors for 4-GHz signals.

Deep Space Communications
Route 1, Box 351A
Chancellor, AL 36316
(205) 393-3211
Complete home earth stations featuring a 12-foot antenna, polar mount, LNA, feed horn, receiver, and all necessary cables.

Delta-Benco-Cascade Ltd.
124 Belfield Road
Rexdale, ON, Canada, M9W1G1
(416) 241-2651
Cable television and other equipment.

Dexcel
2285C Martin Avenue
Santa Clara, CA 95050
(408) 727-9833
Low-noise amplifiers and satellite receivers. Also complete electronics packages (less antenna and mount) from $2,795 to $3,220.

Discom Satellite Systems Inc.
Box 8699
Independence, MO 64054
(816) 836-2828
Antennas for home earth stations.

Downlink Inc.
Box 33
Putnam, CT 06260
(203) 928-7955
Complete spherical and parabolic antenna systems and complete

home earth stations. Systems ranging from $3,595 to $4,595 and receivers from $1,295 to $1,895.

Robert L. Drake Co.
540 Richard Street
Miamisburg, OH 45342
(513) 866-2421
Satellite receivers.

Earth Terminals
255 Northland Boulevard
Cincinnati, OH 45246
(513) 772-6900
Satellite receiver including remote control unit with a 25-foot cable. Options include a feed rotation control. Price is $2,000.

Earthstar Corp.
16012 Cottage Grove
South Holland, IL 60573
(312) 755-5400
Home earth stations and components such as the ETR-2 receiver, FG-3 antenna with polar mount, and the MA-120 LNA.

Echosphere Corp.
5315 South Broadway
Littleton, CO 80120
(303) 797-3231
Complete home earth stations and components.

Energy Systems Limited
2306 Charles Avenue
Dunbar WV 25064
(800) 624-9046
Home earth stations and components, accessories, and cable.

Gardiner Communications Corp.
1980 South Post Oak Road
Suite 240
Houston, TX 77056
(713) 961-7348
A complete line of home earth station products including antennas, LNAs, receivers, Television modulators, and LNA power supplies.

Gillaspie and Associates
950 Benicia Avenue
Sunnyvale, CA 94086
(408) 730-2500
Satellite receivers and converters for home earth stations. The model 7500 receiver has two independent audio channels, LNA theft alarm, built-in modulator, and signal strength carrier level meter. Price is $1,295.

H&G Systems
15109 Chicago Road
Dolton, IL 60419
Home earth station receivers.

H&R Communications Inc.
Route 3, Box 103 G
Pocahontas, AR 72455
Complete home earth stations including 8 to 20-foot antennas, 120-degree LNA, and everything else needed. Prices range from $2,500 to $8,550.

Harrell's Southside Welding
Old Highway 7 North
Route 2, Box 46
Grenada, MS 38901
(601) 226-4081
Earth station antennas.

Hastings Antenna Co.
847 West First Street
Hastings, NE 68901
(402) 463-3598
Earth station antennas

Heath Co.
Benton Harbor, MI 49022
(616) 982-3210
Series of kits to build a complete home earth station. The sequence starts with a site survey kit for $30 and proceeds from there with a mix of kits and ready-built products. The installation features a Scientific-Atlanta antenna, a 24-channel receiver kit, and a LNA with a motorized feed assembly. Total price of the package is around $6,900.

Helfer's Antenna Service
23 Brookside Place
Pleasantville, NY 10570
(914) 769-2588
Complete home earth stations and components.

Hero Communications
1783 West 32nd Place
Hialeah, FL 33012
(305) 887-3203
Complete earth stations from $7,100 to $15,000. Systems feature 12 to 20-foot parabolic antennas on polar mounts and include all necessary cables, etc.

Houston Satellite Systems Inc.
8000 Harwin, Suite 397
Houston, TX 77036
(713) 784-8953
Tracker III antenna controller. Features a built-in microprocessor to give positive, precise antenna pointing. Includes automatic overshoot correction and preprogrammed error codes to indicate improper satellite selection and actuator and rotor malfunctions.

Howard Engineering
Box 48
San Andreas, CA 95249
Circuit boards to build satellite receiver.

Hughes Corp. Microwave Communications Products
Box 2999
Torrance, CA 90509
(213) 534-2146
Complete home earth stations.

Hustler
3275 North B Avenue
Kissimmee, FL 32741
(305) 348-5111
Complete earth stations featuring 3-meter dish, 120-degree LNA, and 24-channel receiver with a built-in modulator.

Industrial Scientific Inc.
3280 Wynn Road
Las Vegas, NV 89102
(702) 362-2405
Northstar antennas with precision pointing capability and a steel mesh surface. Polar mount.

International Crystal Manufacturing
10 North Lee
Oklahoma City, OK 73102
(405) 236-3741
Satellite receiver. The ICM TV-4400 receiver features a separate down converter for mounting at the LNA. Price is $1,395.

International Video Communications
4005 Landski Road
North Little Rock, AR 72204
(501) 771-2800
Home earth stations for about $4,000.

Intersat Corp.
2 Hood Drive
St. Peters, MO 63376
(800) 325-6122
Complete home earth stations and satellite receivers.

Interstar Satellite Systems
21708 Marilla Street
Chatsworth, CA 91311
(213) 882-6770
Complete home earth stations and equipment.

KLM Electronics Inc.
1170025 Laurel Road
Morgan Hill, CA 95037
(408) 779-7363
Complete home earth stations including their own brand of receivers—Sky Eye antennas. Earth stations start at $5,990.

Kosmos Systems Inc.
11985 U.S. Highway One, St. 207
Juno, FL 33408
(305) 626-3800
Antennas.

Leaming Industries
180 McCormick Avenue
Costa Mesa, CA 92626
(714) 979-4511
FM stereo processors.

Lindsay America
Consumer Electronics Division
2608 East Hills Drive
Williamsport, PA 17701
(717) 326-7133
Satellite antennas in three sizes: 8, 10, 12 and 15 feet. Also available are feed horns, television modulators, and complete earth stations. Complete earth stations range from $5,700 to $10,000.

Lowrance Electronics
12000 Skelly Drive
Tulsa, OK 74128
(800) 331-4105
Satellite receivers.

Robert Luly Associates
Box 2311
San Bernardino, CA 92405
(714) 888-7525
Umbrella-type lightweight antenna uses mylar stretched on ribs for a reflecting surface. Complete systems.

Mac Line Inc.
West 3125 Seltice Boulevard
Couer d' Alene, ID 83814
(208) 765-0909
Antennas and complete home earth stations. Antenna line includes a 12-foot one-piece parabolic antenna, a 12-foot parabolic antenna in seven pieces, and a rectangular parabolic antenna.

McCullough Satellite Systems Inc.
Box 57
Salem, AR 72576
(501) 895-3176
Spherical antennas including the 8-Ball antenna kit for home earth stations.

Merrimac Industries Inc.
41 Fairfield Place
West Caldwell, NJ 07006
(201) 575-1300
Receivers and down converters.

Microdyne Corp.
Box 7213
Ocala, FL 32672
(904) 687-4633
Complete home earth station featuring a 12-month warranty. Cost is about $10,000 with a 12-foot antenna and about $11,500 with a 4-meter dish.

Micro-Link Systems Inc.
6012 Dixie Drive
Dayton, OH 45414
(201) 575-1300
Receivers and down converters.

M/A COM (Microwave Associates Communication)
63 Third Avenue
Burlington, MA 01803
Complete home earth station. The Series Four earth station features a 10-foot antenna with motor drive, two 120-degree LNAs, and the Series Four receiver. Price of the complete system is $8,450.

Microwave Filter Co.
6743 Kinne Street
East Syracuse, NY 13057
(800) 448-1666
Filters for earth stations.

Microwave General
2680 Bayshore Frontage Road
Mountain View, CA 90019
Complete home earth stations. The Star Trac system costs $5,995. Options include 100-degree or 80-degree LNAs for weak signal areas.

Mid-America Video
324 Pershing Boulevard
North Little Rock, AR 72204
(501) 753-3555
Complete home earth stations. The Galaxscan system sells for $3,349 with a 10-foot antenna and $3,849 with a 13-foot antenna.

Mini-Casat
Route 3, Box 150
Fulton, MS 38843
(601) 862-2132
Antennas and satellite receivers.

Muntz Electronics Inc.
7700 Densmore
Van Nuys, CA 91408
(213) 782-7511
Complete home earth station featuring 13-foot dish, azimuth mount, 120-degree LNA, and receiver for $3,995.

National Microtech
Box 417
Grenada, MS 38901
(800) 647-6144
The Apollo line of home earth stations including the distinctive square-parabolic antenna. Prices range from $6,800 to $8,800 depending on choice of LNA.

National Satellite Communications
Plaza 7
Latham, NY 12110
(518) 783-0088
Distributors of many brands of earth station components including ADM, Auto-Tech, Avantek, AVCOM, Chapparal, Drake, Dexcel, Comtech, Luly, Pilot-View, Sat-Tec, Standard, Vector, Vidare, and Washburn.

Newton Electronics
2218 Old Middlefield Way
Mountain View, CA 94043
(415) 967-1473
Satellite receivers.

Norsat Systems
Box 1234
Blaine, WA 98230
(604) 585-9428
Two-part satellite receiver. Front end built into LNA and down converter and demodulator in unit that goes in the house.

North Supply Co.
10951 Lakeview Avenue
Lenexa, KS 66219
(913) 888-9800
Antennas and satellite receivers.

Orlando Antenna Co.
2356 West Oakridge Road
Orlando, FL 32809
(305) 851-8332
Satellite antennas.

Paraframe Inc.
Box 423
Monee, IL 60449
Large antenna kits and antenna design manuals.

Pinzone Communications Products, Inc.
10142 Fairmount Road
Newbury, OH 44065
(304) 296-4493
Complete home earth stations and satellite receivers. The Pinzone 8200 satellite receiver features automatic fine tuning, dual LNA input, and a variety of options.

Prodelin Inc. (a division of MA/COM)
Box 131
Hightstown, NJ 08520
(609) 448-2800
Antennas for home earth stations in sizes ranging from 6 to 15 feet.

Quantum Associates Inc.
Box 18
Alpine, WY 83128
(307) 654-2000
Remote antenna positioners.

Rohner & Associates
501 North Elm Street
West Liberty, IA 52776
(319) 627-4819
LNA kits and receivers. The receiver comes assembled and tested and features all-in-one tuning and a built-in power supply. Price is $900.

Sales Inc.
688D Alpha Park
Cleveland, OH 44143
(216) 461-0000
Home earth stations and components. Carries Gillaspie, Standard Agile, KLM, and Drake receivers and Astra antennas.

Satcom Canada
1140 Claire Crescent
Montreal, PQ Canada, H8S1A1
Home earth station antennas and receivers. The antenna is a 12-foot aluminum model on a polar mount with an optional expansion kit to increase its diameter to 15 feet. The receiver features digital push-button tuning with auto-scan and a programmable timer-videotape recorder interface.

Satelco
5540 Pico Boulevard
Los Angeles, CA 90019
(213) 931-6274
Home earth station antennas and other components.

Satelinc
Riverside Drive
Chestertown, NY 12817
(518) 494-4151
Complete home earth stations.

Satellite Communications Inc.
8101 Gallia Street
Wheelersburg, OH 45694
(614) 574-8121
Complete home earth stations.

Satellite Earth Station Technology
4369 Creditview Road
Mississauga, ON, Canada, L5M2B5
(416) 826-8066
Home earth station antennas in sizes ranging from 8 to 12 feet. They come with wireless control and cost between $3,500 and $4,500.

Satellite Reception Systems, Inc.
2370 Morse Road
Columbus, OH 43229
(614) 471-6118
Earth station components including receivers by AVCOM, Drake, Dexcel, Auto-Tech, KLM, Sat-Tec, Comtech, Washburn, M/A-COM, and Microdyne. Antennas by Luly as well as the company's own brand and polar mounts.

Satellite Supplies
Box 278
Aldergrove, BC, Canada, VOX1A0
Satellite receivers featuring the down converter on the antenna.

Satellite Systems Unlimited
Box 43
Conway, AR 72032
(501) 327-6501
Satellite receivers with the down converter mounted on the antenna.

Satellite Technology Services
11684 Liburn
St. Louis, MO 63141
(800) 325-4058
Satellite receivers at $1,495 and $2,000.

Satellite Television Systems
Box 22557, Suite 1140
Denver, CO 80222
(303) 750-0564
Complete home earth stations from $3,995 to $12,000.

Satellite Television Systems
Box 51837
Lafayette, LA 70505
Home earth stations and antennas.

Satellite Television Technology International, Inc.
Box G
Arcadia, OK 73007
(405) 396-2574
Satellite receivers.

Satfinder Systems Inc.
6541 East 40th Street
Tulsa, OK 74145
(718) 664-4466
Complete home earth stations.

SAT Share Inc.
5301 Hollister, Suite 100
Houston, TX 77040
(713) 460-9900
Complete home earth stations.

Sat-Tec Systems
2575 Baird Road
Penfield, NY 14526
(716) 586-3950
R2B satellite receiver featuring variable audio tuning to give a complete selection of subcarriers. Built-in power supply can handle two LNAs.

SAT Vision
Box 1490
Miami, OK 74354
(918) 542-1616
Satellite antennas.

Scientific-Atlanta Inc.
3845 Pleasantville Road
Atlanta, GA 30340
(404) 449-2000
A major manufacturer of antennas for home earth stations and other microwave uses. Also LNAs and other equipment.

Scientific Communications Inc.
3425 Kingsley Road
Garland, TX 75041
(214) 271-3685
Satellite receivers and LNAs.

SED Systems Ltd.
2414 Coyle Avenue
Saskatoon, SK, Canada, 57K3P7
(306) 664-1825
Antennas and complete home earth stations.

Simcomm Labs
Box 60
Kersey, Co 80644
(303) 352-1020
LNA for home earth stations.

Skyscan Corp.
250 East 36th Street
Tucson, AZ 85713
(602) 622-2261
Complete home earth stations featuring a 10-foot antenna. Prices start at $3,995.

Southern Fiberglass Supply
1105 Tuckahoe Drive
Nashville, TN 37207
(615) 868-4094
Antennas and mounts. Both polar and EL-AL mounts are available. The company's 13-foot dish sells for $1,195.

SRC Industries
773 South Oregon
Ontario, OR 97914
(503) 889-7261
Home earth station antennas.

Standard Communications
Box 92151
Los Angeles, CA 90009
(213) 532-5300
Satellite receivers

Star Trak Systems
404 Arrawanna Street
Colorado Springs, Co 80909
(303) 475-7050
Satellite receivers and antennas.

Star Vision Systems
611 North Wymore Road
Winterpark, FL 32789
(305) 628-5458
Complete home earth stations.

SWAN TVRO Satellite Systems
614 Cimarron
Stockton, CA 95210
(209) 478-2756
Antenna kits and complete home earth stations. Antennas come in 12-foot, 14-foot, and 5-meter sizes. The 12-foot kit sells for $1,075, the 14-foot kit is $1,800, and the 5-meter kit is $2,000.

Telecom Industries Corp.
27 Bonaventura Drive
San Jose, CA 95134
(408) 262-3100
Satellite receivers.

Tel Vi Communications
1307 West Lark Industrial Boulevard
St. Louis, MO 63026
(314) 343-9977
Antenna drive systems. Lead screw design with antijam clutch.

Third Wave Communications Corp.
2600 Gladstone
Ann Arbor, MI 48104
(313) 996-1483
Complete home earth stations.

TL Systems
1101 East Chestnut Avenue, Suite C
Santa Ana, CA 92701
(714) 547-1981
Satellite receivers and television modulators.

Transtar Communications Corp.
1408 Columbia Avenue
Franklin, TN 37064
(615) 244-6112
Complete home earth stations.

Transvision
2100 Redwood Highway
Greenbrae, CA 94904
(415) 924-6963
Complete home earth stations. Features 11-foot antenna, 120-degree LNA, remote LNA rotor, and a KLM receiver for $5,995.

Wagner Industries
Box 559
Alva, OK 73717
(405) 327-1877
Eagle spherical antennas in a variety of sizes.

Western Satellite
Box 22959, Wellshire Station
Denver, CO 80222
Complete home earth stations.

Wilson Microwave Systems, Inc.
4286 Polario Avenue
Las Vegas, NV 89103
(800) 624-6898
Complete home earth stations and components.

Chapter 7

The Sweet Sounds of Stereo

Movie theaters have many things going for them that help to bring customers back. Theaters have first-run movies, great popcorn, wide-screen viewing, and tremendous sound.

At home you can have the popcorn, duplicate the wide-screen viewing with a projection television, and get commercial-free, almost first-run movies via satellite. Unfortunately, it has been next to impossible to get much more than the rattling of tin cans from the cheap speakers built into today's television sets.

The home viewing of movies that have great sound tracks is bound to be one of life's great disappointments, right? Wrong. Many programmers are now transmitting true stereo feeds along with their video. There are plenty of audio-only feeds coming off the satellites and just waiting for eager listeners. Imagine for a moment the absolute thrill of having the true-to-life sound of Phantom jets screaming through your living room as you watch *Apocalypse Now.* You can enjoy the crooning of Gene Kelly and Olivia Newton-John when the family is watching *Xanadu.* How about the pleasure of watching a concert and getting the true concert sound in fine stereo?

STEREO BROADCASTING

The stereo signals are principally broadcast to benefit the cable television system operators as an enhancement to their premium movie channel offerings and to help the operators fill many of the available channels. Many audio-only feeds are available to the cable operators at no charge, which makes them an ideal service if an operator has 100 or more channels that have to be filled while still making a reasonable profit.

Not much extra equipment is needed to tune in on stereo signals, especially if you happen to own a home stereo system already. The principal piece of equipment is the stereo processor, which has the job of tuning, filtering, and outputting the stereo signals. Cable television systems need a commercial stereo tuner to provide volume, balance, and tonal control. A home stereo system will work just as well for a home or small commercial setup. Cable television systems and commercial apartment complex-type setups need to add stereo FM generator/modulator units at the head-end of their systems. They need to provide their subscribers with standard FM taps.

TRANSMISSION TECHNIQUES

It seems simple enough on the surface. Pull in the stereo signal, electronically massage it a bit, and you instantly have decent audio. It's not quite that simple, though. The various programmers cannot decide on a standard method of transmitting stereo signals (commercial broadcasters have the same problems with the newly allowed AM stereo). It's not an easy task getting stereo. When stereo signals were first being sent, all transmissions were multiplexed, audio and video together, and transmitted in a manner similar to commercial stereo FM broadcasts. Such signals use a signal subcarrier for the audio and are compatible with monaural receivers because the receivers get the full audio signals, whether or not they split the signal into stereo.

A much more popular transmission technique is to use two separate subcarriers for the audio signal. Since August 1,

1981, Warner-Amex has been broadcasting MTV on Satcom F1 transponder 11 using two subcarriers as a means of obtaining better fidelity and signal separation. Unfortunately, if one subcarrier was given one channel and the other subcarrier the opposite channel, monaural receivers would be left with only half a signal. They could only tune a single subcarrier at a time. To eliminate this problem, Warner-Amex uses a matrixed style of transmission wherein one subcarrier is given a signal consisting of the left channel and the right channel combined (called left plus right or L+R). The left channel minus the right channel (L−R) is sent on the other subcarrier. With this technique, a monaural receiver tuned in on the L+R subcarrier will get the full monaural sound.

A stereo processor, however, would bring in both subcarriers simultaneously. By doing some electronic things to the signal, such as inverting the signals and adding the results together, it would actually separate the left channel from the right channel and deliver true stereo signals. This technique is used by The Movie Channel.

CBS Cable transmits its signal in a similar manner. It additionally multiplexes half of the signal—a technique referred to as *adaptive deviation.*

This doesn't mean that you won't find a single channel being broadcast on a single subcarrier. Many firms are doing so. They are adding a third audio subcarrier consisting of the L+R signal, so that those with monaural receivers won't be left out. Times Mirror Corporation uses this technique with its Spotlight channel. Special event uplinks such as concerts use discrete stereo without the third subcarrier.

The transmissions in the previous examples use a very wide deviation, which generally limits a satellite transponder to three audio subcarriers along with a single video carrier. The people who own the various subcarriers obviously want the most signals possible to maximize their revenues. If you check Satcom F3R, you'll likely see an example of a very busy transponder. You'll probably find that the United Video transponder is transmitting WGN-TV's video and wide deviation audio, but you'll also discover that the remaining space has been optimized through the use of numerous narrow deviation discrete stereo formats. Early in 1982 we counted 11 such subcarriers in addition to WGN's audio and video.

BUYING A STEREO PROCESSOR

If you want a stereo processor, don't buy it on the basis of price alone. Not every processor will necessarily be able to process every format signal. Following is a list of features that you should look for in a stereo processor after determining the applicability of each item on the list:

- It should be able to receive and process a single subcarrier that may or may not be multiplexed.
- It should be able to receive and process two separate subcarriers simultaneously. The best units on the market have two separate tuners. There is no agreed-upon standard that determines the difference in frequency and spacing between the individual subcarriers. If such a standard does come into effect and is followed by the broadcasters, only a single tuner will be necessary. The unit will automatically lock onto the second subcarrier.
- It must be able to process L+R and L−R matrixed subcarriers as well as left and right discrete subcarriers.
- Because you want to eliminate unwanted "noise" without losing any of the signal, an excellent tuner will have a variety of selectable noise filters. If a unit only has a single noise filter that is too narrow, it will cut off parts of the signal. If the noise filter is too wide, it will let in too much noise.

A stereo processor needs to get an unfiltered composite video signal in order to work properly. Most standard satellite television receivers deliver this type of signal through one of its normal outputs, although it may be necessary to connect an internal lead to get the signal.

As of this writing, all known stereo processors will output separate left and right channels, which should be input to a stereo tuner or amplifier for volume, balance, and tonal control. A tuner may be avoided if you use one of the new stereo televisions. Such sets may still be difficult to find in the United States, but nearly 4 million sets have been sold in Japan. More than 10 percent of all Japanese households had these sets by mid-1982. With stereo television, the output signal from the stereo processor goes directly to the television. With older, monaural television, the signal must bypass the television's monaural audio. This is done by routing the stereo processor's output signal to a regular stereo tuner and then on to a pair of matched speakers.

MANUFACTURERS OF STEREO PROCESSORS

Following is a list of manufacturers of stereo processors. With this list and any others in this book, there are a few things to remember. This book was researched in 1982, and companies listed may no longer be in business. Conversely, there may be other firms now providing the same type products and services so the list is not necessary all-inclusive. Finally, this is a fairly young industry. Not all firms are reputable, so let "buyer beware" be your watchword.

Arunta Engineering Co.
P.O. Box 15082
Phoenix, AZ 85060

Crystal Manufacturing Co.
10 N. Lee
Oklahoma City, OK 73102

Chapter 8

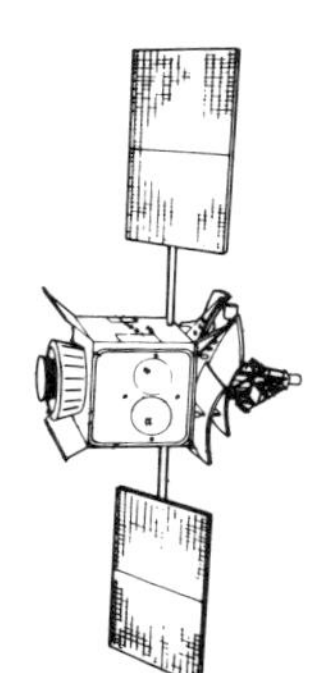

Low-Power Television

Besides commercial broadcast television, cable, multipoint distribution systems, and the proposed direct broadcast systems, another form of broadcast video is about to enter the picture. The newest entrant is LPTV or low-power television.

It costs a lot of money to start up a commercial broadcast outlet in the United States today. To be a major station on the vhf spectrum (channels 2-13), a station has to transmit more than 100,000 watts of power. Stations broadcasting closer to 500,000 watts are not unusual. On the uhf end of the dial, most stations broadcast more than 1 million watts.

LPTV AND THE FCC

The broadcast equipment is obviously quite expensive. The FCC has reams of rules and regulations that dictate how the station must be operated, and the competition is intense. As a result, many minority groups are deprived of the opportunity to participate in the television broadcast industry. To try and ease the crunch and encourage minority participation, the FCC devised LPTV.

A vhf LPTV station will be limited to 100 watts of broadcast power, and a uhf station will be limited to 1,000 watts. With streamlined operating requirements and licensing procedures, the FCC hopes that LPTV stations will proliferate—especially in smaller communities that traditionally could not afford a television outlet of their own.

Because LPTV stations will have a broadcast range of less than 25 miles, the FCC hopes that these stations will serve local community needs and will help identify and solidify smaller communities. One key to this will be that the FCC doesn't appear to have any plans to restrict the profitability of LPTV stations, so commercials are bound to be numerous.

When the FCC first announced plans to accept applications for LPTV outlets, they were swamped with 5,000 requests. Unfortunately, most applications were for stations within the top 80 to 100 existing markets. Most of the potential station owners wanted rights for the most potentially profitable areas of the country. According to most of the applications on file, operators plan to rely on minority, cultural, and religious programming. A number of operators plan to use programming available on satellite. The LPTV stations are allowed to originate their own programming, so future viewers may be in for a potpourri of programming with widely varying quality.

LPTV AND CABLE TELEVISION

The LPTV option may impact cable television systems, especially in the rural areas, since the start-up costs are only expected to be about $25,000, as compared with an estimated $2 million for a commercial television station. The low start-up expense, plus the unrestricted profit potential, may sway some cable operators who are looking for a faster return on their investment. Cable operators may even choose to get in on LPTV to become established in a community or to help finance cable construction.

Another advantage of LPTV over cable is that local communities haven't shown an interest in trying to regulate LPTV. They do not see it as a potential revenue source for the municipal coffers.

The first of the LPTV stations began operating early in 1982. The FCC has tightly controlled the rate of start-up so they can change the rule of the game if necessary, we suspect. There is little doubt that LPTV will become a major broadcast force in this country.

The presence or absence of LPTV outlets in a community should have little or no impact on privately-owned earth stations. The programming available will probably be of little or no interest to someone who is seriously considering the benefits of an earth station. It is possible, however, that a local LPTV outlet will become one of the offerings on a local cable firm's basic cable.

Chapter 9

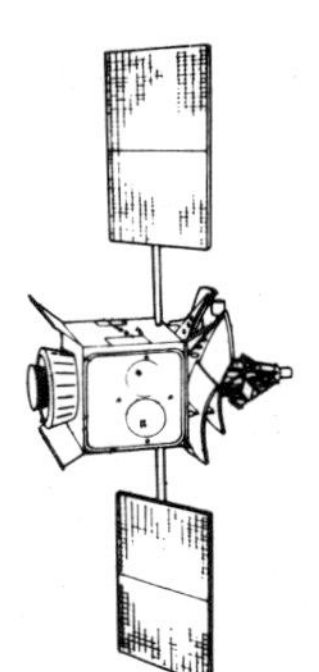

Cable Television

Other than the world's telephone companies, there is perhaps no industry except for the domestic cable television industry that owes its booming success to existence and heavy use of communications satellites (Fig. 9-1).

According to the National Cable Television Association (NCTA), the past decade has brought phenomenal acceptance and growth to cable television. In 1968 some 2,000 cable television systems were wired into a mere 2.3 million of the nation's 56.3 million television homes. By January 1 the number of systems had more than doubled to 4,300. Those systems were serving more than 19 million television households and growing rapidly. A significant portion of that growth must be attributed to the wide variety of programming currently available on satellites—programming that can be sold for a handsome profit.

Cable television had its beginnings after World War II when television became the rage but was plagued by poor reception caused by geographic problems. People living in valleys often received signals that had bounced off the valley walls several times, resulting in broken-up images. Mountains and tall buildings interfered with reception. Some major population pockets were stranded with poor, little, or no television reception.

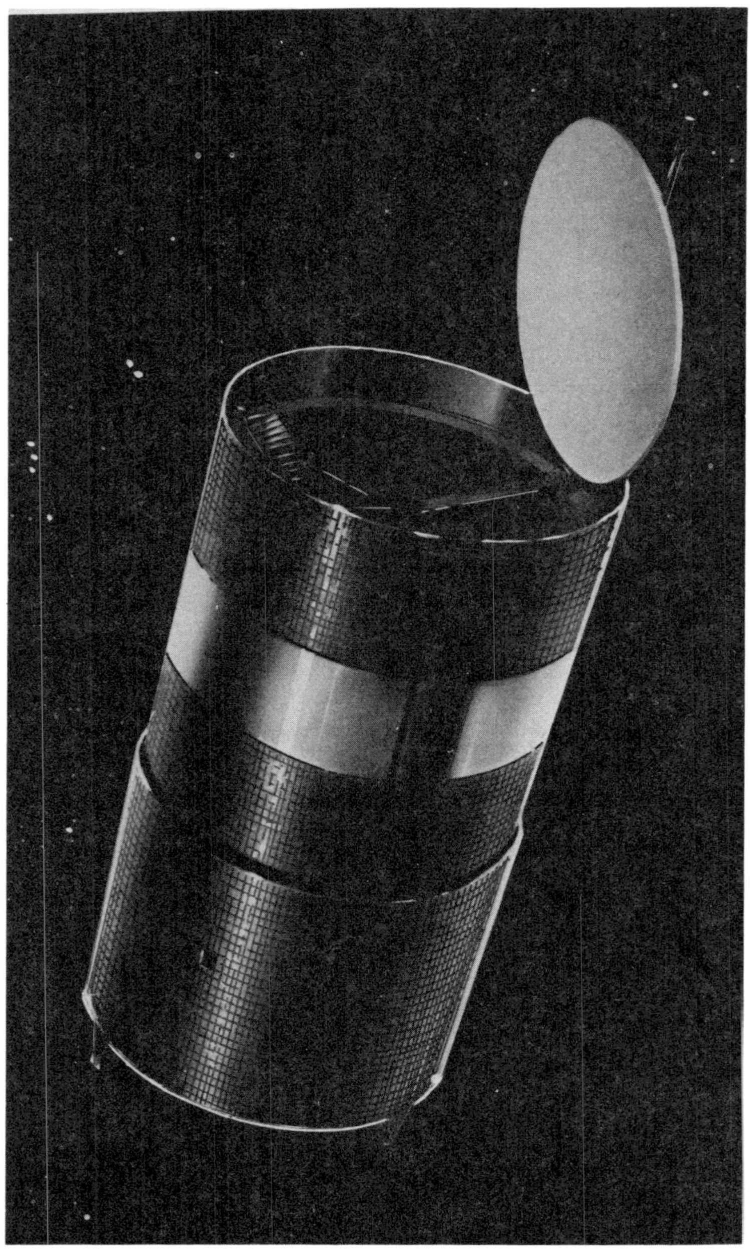

Fig. 9-1. Westar IV has the capacity to handle 24 channels of television simultaneously.

COMMUNITY ANTENNA TELEVISION (CATV)

America had just returned victorious from a world at war. It would have been unthinkable not to use advanced technology to help provide Americans with their well-deserved entertainment. So Community Antenna Television (CATV) was born—a master antenna system for communities with little or no television reception. The CATV systems usually carried a small handful of local stations. The service brought television to entertainment-starved Americans for less than $5 a month.

The CATV industry was off and running—or so it seemed. In 1962 a company called the Carter Mountain Corporation had an application on file with the FCC for permission to use a microwave system to beam signals into a community that already had one local television station. In what the NCTA calls a landmark decision, the Federal Communications Commission decided that if a CATV company brought an outside signal into a community that already had a television station, that station faced sure economic ruin. Carter Mountain's application was denied, and the FCC took jurisdiction over all microwave-fed CATV systems. In 1966 the FCC expanded its reach to rule over all cable television systems.

FCC RULES

In 1968 the FCC decided to make cable television more interesting by imposing a number of "rules" that cable systems had to abide by. If cable operators wanted to bring in a distant station, they had to receive permission on a program-by-program basis from the originating station. Additional permission were required if the cable systems were within 35 miles of the top 100 television markets and wanted to import a signal. The cable systems that were within 35 miles of a station not located in a top 100 market didn't need special permission to bring in the three major networks and one independent station.

The "luckiest" cable operators were those located in communities more than 35 miles from any commercial television station. Those folks could bring in as many distant signals as they wished.

In the dark ages before Ted Turner and his superstation concept, the cable companies had to bring in the closest independent station if they brought any in at all. Viewers would be given programming based on proximity, not quality.

Not content with their 1968 work, The FCC decided to "help" the cable television industry with some rules covering signal carriage, access and nonbroadcast activities, technical standards, and federal/state/local relationships. The FCC also bestowed on the industry the "certificate of compliance," which became a precondition for cable television system operation.

Many 1972 rules and regulations were later tossed out by various courts as overrestrictive or overregulation by the FCC. As the courts began tossing out many FCC-imposed rules and regulations, the cable television industry began to change its shape, services, and growth to what we have now. The FCC has begun to de-emphasize its regulation of the industry. The cable television industry is refocusing its fights to state level, where totally different battles are being waged.

PROTECTION

Author Cliff Page of the National League of Cities Urban Features Syndicate says the big issue in cable television these days is "protection." City officials want to protect their authority to regulate cable television in order to protect consumers—local citizens—from a business that is by nature a monopoly. They want to ensure the widest possible competition.

Cable companies want the constitutional protection of the First Amendment and the legislative protection of antitrust law to shelter them from all regulation. Both sides want to wire the nation's cities with coaxial cable television, but each side wants to protect its stake in an industry that generates large amounts of money. At the heart of the debate is the question of what cable television is: an entertainment medium, an information system, or an enhanced version of something we already have.

CABLE TELEVISION AND CITIES

Cable television is deeply rooted in Seattle, a city whose spectacular waterfront is abutted by hilly terrain that interferes with normal television reception. The community has witnessed the evolution from retransmission of local television programs into the satellite age of modern technology. Along the way, Page writes, local officials developed a keen understanding of cable issues.

"To the average citizen, cable offers the excitement and choice we've expected from television since the beginning," says Mayor Charles Royer. "To local officials it also is an extraordinary communications system that can serve the entire community in a way that is unique and unrivaled by anything else."

Royer, who heads a special National League of Cities task force on cable television, sees many new public policy issues accompanying the great diversity in programming and services that are the hallmark of new urban cable systems.

The prospect of treating cable as a common carrier—like the telephone company—is an idea that "horrifies the cable industry," Royer says. On the other hand, he adds, the industry's insistence on being treated as an "electronic newspaper . . . ignores the realities of the marketplace" and thinly disguises a deep-rooted effort to achieve deregulation.

CABLE TELEVISION FUNCTIONS

Cable can best be understood and dealt with if regarded as having two distinct identities—a transmitter and a programmer—says Nicholas Miller, a communications lawyer in Washington, D.C. The key to regulation of cable television, Miller believes, is to understand the difference between the two functions.

"A cable company is really two separate companies," says Miller. "One builds, operates, and maintains the system. It's a natural monopoly, comparable to the telephone company. The other is a programming broker, totally separate from the physical plant. It selects and offers video, teletext, security, consumer, and other services."

The transmission system is inherently monopolistic Cable systems would not be economically feasible if more than one wire ran past each house.

Miller says the programming function "should face real market competition." He sees cable companies willing to use their monopoly to "maximize the profits from programming by limiting or denying access to the system." This, he says, leads to price discrimination and to cross subsidization of systems.

For instance, Page writes, a cable operator could use a portion of his revenues from subscriber fees for movies and sports coverage to undercut the price of anyone who might attempt to lease a channel and offer a home security service or catalog service that would compete with a similar service offered by the cable operator.

REGULATION

As Mayor Royer pointed out at a National League of Cities meeting, some companies will use an older, profitable cable system to subsidize their efforts to win new franchises by underpricing the true cost of the newer, more complex systems. Royer described the practice as "turning one city into a cash cow," giving capital for construction in another town.

It is partly because of these practices—or cable television's "failure in the marketplace," as Royer puts it—and partly because the public sees a need for "someone to help make sure that an industry serves the public," that there is pressure for regulation of cable television.

Rate regulation by local government, says Royer, serves to keep a city from being used as a "cash cow" by requiring reinvestment in the system generating the cash. A major purpose of such reinvestment is to develop and support local access to a cable system to preserve the public's right to hear a broad range of speakers.

"The industry needs to understand that it is local government that citizens turn to for this kind of protection," Royer says.

Cable operators generally have a different view. Harold Farrow, an Oakland, California, attorney who has represented the cable industry, says cable regulation evolved "not to protect the citizens, but to protect the television stations."

According to Farrow, cities claim responsibility to regulate, "but the documents they write aren't regulations or laws. They all require the acceptance of the operators. You are trying to accomplish by contract what you don't have the police power to do through regulation."

The cable industry has twice launched all-out lobbying campaigns to push cable deregulation through Congress. Both times local officials led by the National League of Cities preserved the authority they now have to negotiate franchises and regulate basic rates. (Fees for special cable services such as movies are not regulated.)

For their part, says Farrow, cable operators believe they can accomplish—through competition and without regulation—the same things that are of concern to local governments.

Though cities were successful in maintaining a role in cable regulation, Royer cautions that there is a delicate balance which must be struck between local leaders and cable operators. "Just as the cable companies have overpromised (about services and access), we've overasked. We need to let the marketplace work by stating basic needs and finding out how the companies will meet them."

Taking a totally different viewpoint is Thomas E. Wheeler, president of the National Cable Television Association. In 1981 he told a congressional subcommittee that, "Cities have used the franchise process to extend local regulation beyond its proper scope into virtually every aspect of a cable system's operation."

Referring to local government control of cable as "ill founded," Wheeler emphasized that "congressional scrutiny and appropriate federal limits on municipal government involvement in franchising are necessary to prevent the potential subversion of national communications policy objectives for irrelevant and inappropriate reasons." He appeared before the House Subcommittee on Telecommunications, Consumer Protection, and Finance.

Wheeler listed "excessive franchise fees, unjustified rate regulation, access channels for each and every political constituency and . . . actual ownership and content control of the cable system" as government intrusion into media that poses serious problems. "They misconstrue the proper role of local cable regulation and seize upon legitimate municipal concerns as a pretext to extend government involvement far beyond its proper scope."

Wheeler called for a federal ceiling on franchise fees, basic rate deregulation where alternatives to cable television are reasonably available, a federal prohibition against governmentally mandated access to cable television systems, and a federal prohibition of government ownership of cable systems.

FRANCHISE FEES

A franchise fee ceiling, Wheeler said, "focuses the competition on service innovation to benefit the consumer rather than on bidding up the level of monetary payment made to the city." He mentioned that cities squeezed by federal aid cutbacks and post Proposition 13-type budgets are viewing cable television franchising as a potential solution for their financial woes.

"The forces of competition and financial need can make franchising an extremely volatile process: cable companies competing for the franchise to wire a city are at the mercy of local officials, who, in turn, are faced with mounting deficits and diminishing revenues," Wheeler noted.

The federal interest in continuing to limit franchise fees, he said, is related to the risk that excessive franchise fees will deprive the cable industry of needed revenues for developing as a nationwide network. Referring to the interest of current and potential cable television consumers, he pointed out that "excessive franchise fees are a hidden tax applicable only to cable subscribers."

Wheeler said that rate regulation of cable television is unjustified with one exception—an area that would have no television reception without cable. Cable television "cannot be characterized as a public utility and therefore should not be regulated like one," Wheeler noted. He emphasized that cable fails the utility test on two counts. Cable lacks monopoly control of its market and does not provide essential services for which there are no other alternatives. Recalling that public access was a governmentally imposed response to a legitimate desire for local programming, Wheeler said that the volume of local programming today makes governmentally imposed access unnecessary.

"Our intent is to protect the First Amendment rights and the rights of our subscribers to be free of governmental interference or dictation of program content," he noted. "It does not trample on First Amendment rights for cable operators and cities to jointly agree on programming options as a part of the franchising negotiations. It is contrary to the First Amendment, however, for government to unilaterally impose its programming will on a cable operator."

Wheeler also emphasized that mandated access imposes common carrier-type regulation on cable systems. Also, the cable industry opposed any regulatory requirement that could likely be "a first step toward establishing cable as a separated common carrier and regulating its leased channel rates on a rate-of-return basis."

GOVERNMENT OWNERSHIP

In supporting federal legislation prohibiting government ownership of cable systems, Wheeler said that the issue of freedom of expression should be examined closely. "City officials see it (cable) as not much different from a governmentally owned electric utility. Its role as a part of the nation's communications network is either unappreciated or conveniently ignored."

Municipal ownership of cable presents enormous potential for abuse, Wheeler noted. Financing such a system may prove difficult as well, in light of the enormous revenues needed to build a major urban system and the risk factor in providing a service some may not want, he said.

The battles over cable television are not simply limited

to franchising squabbles between cable operators and the cities. Cable systems often find themselves under the attack of state legislatures that wish to control programming content, public utility commissions that want to regulate the rental rates the cable systems pay utilities for the use of space on utility poles, and citizens' groups that want to control everything imaginable.

PIRACY

Beyond having people who want to run their businesses for them, perhaps one of the biggest problems facing cable systems is that of piracy.

In this era of cheap electronics, even a mediocre electronics hobbyist can easily build a box that will intercept a cable system's signals. Sometimes the theft is as simple as tapping into a buried cable. Other thefts are significantly more complex—especially when the hobbyist is stealing from a multipoint distribution system or is dealing with scrambled premium channels.

Regardless of how difficult stealing a cable's signal might become, it remains significantly less expensive then investing in an earth station and pulling in the signals directly off one of the birds. It is generally illegal to steal a cable system's signal, but that fact doesn't dissuade everyone. Some people intercept cable signals simply for the challenge.

PROGRAMMING COSTS

Most cable systems in the country today offer a variety of programming. The majority of programming is delivered by satellite. The cable systems often divide their offerings into two categories—basic service and premium service.

According to the NCTA, the average cable system's monthly charge for basic service ranges from $7 to $10, depending upon the number of channels provided and the location of the system. The basic service usually consists of the unadulterated retransmission of all or most of the local television stations's programming, plus the signals from one or more of the superstations that operate out of Atlanta, New York, and Chicago. Other basic services usually include news, sports and weather channels, local government access channels, cultural programming, and religious networks.

Most cable systems require all subscribers to take basic cable and then offer premium channels as add-ons. Premium programming usually consists of offerings such as Home Box Office, Showtime, and The Movie Channel. Some cable systems, such as those owned by Times Mirror Corporation have their own proprietary premium movie channels like Spotlight. Cable systems generally charge $7 to $10 per premium channel, but they often group special packages of three or more channels at a special rate.

With basic service, several premium channels, remote control, and more than one wired-in television in a home, the monthly cable bill can easily exceed $40. The December 1981 edition of the *Pay TV Newsletter* estimated that the typical one-time installation fee for cable is $16, the average basic fee is $8.10, and the average pay cable fee is $9.02.

What does the programming cost the typical cable company? According to the January 18, 1982, edition of *CableVision Plus*, here are some of the fees that a cable system pays for programming.

WTBS. A subscriber fee of 10 cents per month (a 20 percent discount with annual prepayment), a 25 cents per subscriber launch assistance fee ($1,000 minimum, a $7,500 maximum), and a maintenance assistance fee of 15 cents per subscriber year (during three promotion periods).

CNN. A per-subscriber fee of 20 cents a month (15 cents if a WTBS subscriber) and a launch assistance fee of 30 cents per subscriber ($10,000 maximum).

MTV. No subscriber fee but a 10 cent per home passed launch assistance fee that applies to charter systems only.

Trinity. The only fee is a 50-50 split on the cost of co-op ads and cable systems, with more than 5,000 subscribers getting a free earth station when they sign up.

POPULARITY OF CABLE

According to A. C. Nielsen figure in November of 1981, 23.2 million of the nation's 81.5 million television households are wired to cable. Arbitron put the subscriber figure at 20.3 million for the same month, and in December of 1981 Paul Kagan Associates, Inc. claimed 23.6 million homes were wired out of the 46 million homes that have been "passed" by cable.

In its January 4, 1982 issue, *Television Digest* reported that there were 4,697 cable systems operating in 12,909 communities in the United States. They showed the growing popularity and almost surprising penetration of cable in America.

The rapid, continuing growth of cable will probably limit the number of privately-owned earth stations in the urban areas of the country, especially while it takes three or more years to get a payback on an earth station when compared to cable. Forces that will continue to create a demand for earth stations in urban areas include systems with above-market pricing, systems with limited offerings, and communities where the "prevailing community standards" will limit adult programming or lead to censorship of R-rated movies.

With the extremely high cost per mile of installing cable and the lack of population density in rural areas, the cable systems are going to concentrate on the cities. That's why the use of earth stations is going to boom wherever there is an absence of cable.

Chapter 10

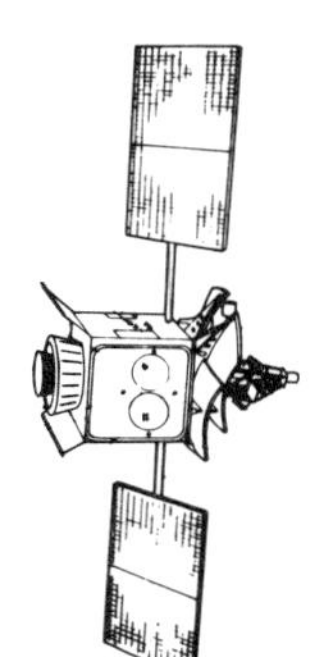

Direct Broadcast Satellite Television

What you pick up on your home earth station is technically not direct broadcast satellite television (DBS-TV, or just DBS). Granted, it's coming direct from the satellite to your house, but you are not the broadcaster's primary market. That signal was really aimed at a distributor—something like a cable television station that would edit it and refeed it to people's houses.

True DBS is aimed at the home. A DBS system would feed a signal from the satellite right to a home antenna. In effect DBS would become a nationwide television channel or series of channels. And that's one of its problems.

Theoretically, you could start broadcasting television programming directly from satellites to the home tomorrow. It is probably going to be the end of the decade before DBS becomes widespread. There are powerful national and international forces that are working hard to see that it never does.

Perhaps more than any other aspect of satellite television, DBS systems epitomize the tangled skein of politics and technology that surrounds the subject. A little background is needed.

THEORY OF DBS-TV

It is perfectly feasible for the average homeowner to set

up a satellite receiving station in his backyard. A fair amount of money and work are involved.

Because the signal is so weak, you need a very large antenna—at least it's large compared to a conventional television antenna. That antenna takes up a lot of room, and it tends to scare people by making them think that satellite television is more complicated than it really is.

If the signal from the satellite was stronger, you could get away with a smaller antenna. Right now the satellite's transponders are only putting out about 5 watts each. If that amount is doubled, the size of the antenna you need shrinks (not in half, because what you are concerned about is the area of the antenna, not its diameter). If the signal is made strong enough, you can shrink the antenna to something about the size of a conventional rooftop television antenna.

Suppose you were only interested in the signal from one satellite. You could do away with the movable antenna mounting, the rotation system, and all the other gear. You could also be more flexible in your choice of a mounting location, because you would only have to see one patch of the sky.

You would have a system that is no more complicated than erecting a conventional television antenna to receive a signal in a fringe area. Many of the problems associated with satellite television would disappear. A lot of freedom of choice would disappear, too, but you'd still have more choice than you do from local stations. That, in a nutshell, is the idea behind DBS television.

The people who are backing the idea are proposing using transponders with 185 watts of broadcast power each. Given that much power, the antenna doesn't have to be more than 2 or 3 feet in diameter. The satellites would use the Ku band (around 12 GHz) on the downlink, a relatively interference-free part of the spectrum.

The idea has attracted considerable international attention. The Japanese and Europeans are both ahead of us in making DBS a reality.

The Japanese were the first to try DBS on an experimental basis. Using a satellite called Yuri (cherry blossom), the Japanese began to test DBS broadcasts in 1981. Testing came to a premature end when the transmitting tubes on the satellite failed.

Undaunted, the Japanese space agency, NASDA, still plans to launch its BS-II communications satellite in 1984 to begin supplying regular DBS service. Meanwhile, the Japanese are continuing their tests with the receivers by using them to receive signals from ground stations.

NHK, the Japanese national television company, is supposed to provide television to all Japan. In some areas conventional television reception isn't possible because of mountainous terrain, or because the signal is blocked by buildings. To provide service in these areas, NHK, in cooperation with NTT, the Japanese phone company, has mounted transmitters on towers to beam signals to the DBS receivers in areas that otherwise would not have television.

The Japanese are also moving aggressively to develop the technology needed to mass-produce DBS receivers. They are developing two kinds of receivers simultaneously. This is unusual for Japan, as companies there have traditionally cooperated closely on basic research. The Japanese attach much importance to the project.

The Europeans have also been working hard to develop DBS systems. Philips, a giant Dutch company that is a leader in technology, has DBS projects going on in Holland, Britain, and France. The French electronics firm of Thomson-CSF has transferred its DBS work from its experimental electronics group to the subsidiary that makes televisions and other consumer electronic devices. The Plessey company in England is working on DBS, and several major German firms have projects going.

Even the Canadians are pushing DBS. The Canadians may put a DBS service into operation this year using their already-launched Anik-C satellite. The method they are proposing to use isn't nearly as elegant as the ones being considered by other countries, and the picture quality won't be nearly as good. But it will give Canadians in many rural areas far better television than what they are receiving now (Fig. 10-1).

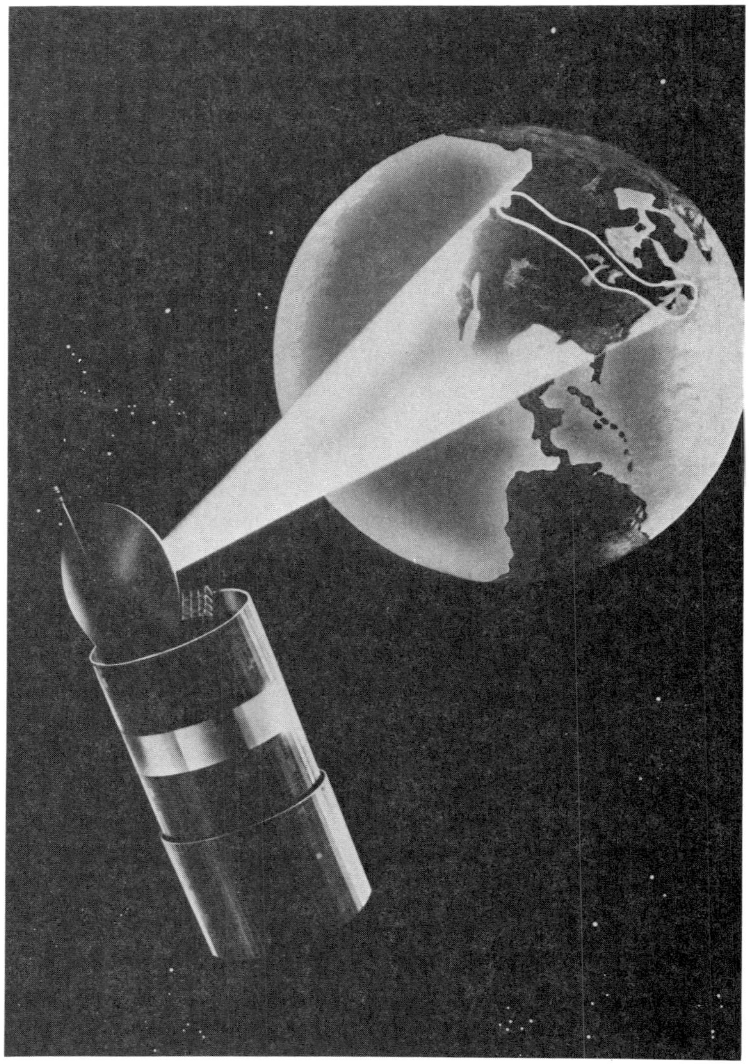

Fig. 10-1. The Anik-D satellites will be focused on Canada (courtesy Hughes Aircraft Company).

THE AMERICAN EFFORT

American companies aren't exactly ignoring DBS, but they aren't pushing as hard as their foreign counterparts. One company that has shown a strong interest in DBS systems is

Scientific Atlanta. Other companies have the technical expertise and are hard at work in fields that have DBS applications, but critical uncertainties are making them leery about pushing for DBS.

If American manufacturers aren't in a rush for DBS, at least some broadcasters are. Widespread DBS television would be a gold mine for the operators. A broadcaster could establish a national network without needing a local affiliate. The broadcaster gets to keep the share of the revenue that usually goes to the local stations. Also, a DBS channel would cover the entire country. There would be essentially no fringe areas, dead spots, or markets without an affiliate.

The FCC started considering regulations for DBS in the United States in 1980. The first company to file a detailed proposal was Satellite Television Corporation, a subsidiary of Comsat Corporation. Satellite Television Corporation was quickly followed by Direct Broadcast Satellite Corporation, a subsidiary of Satellite Systems Engineering, and by Hubbard Broadcasting of St. Paul, Minnesota.

When the FCC began issuing rules in the summer of 1982, there were nine applications pending, including applications from CBS Inc. and RCA, which owns the ABC network. The commission only issued interim rules and did not assign frequencies for the services. That will have to wait until after the Regional Administrative Radio Conference in Buenos Aires later this year (1983).

Licenses can be issued for a maximum of five years under the rules. This is significant, because no one expects to have a DBS system operating in the United States before 1985 at the earliest. Assuming that licenses are issued in 1983, the license period will be half over before broadcasting starts.

In issuing the rules FCC Chairman Mark S. Fowler said they represented "neither a red flag nor a checkered flag for the new video system of direct broadcast satellites. It is simply a green light to a new player."

WORLD SCHEDULE FOR DBS SYSTEMS

Just how green that light is poses a significant question right now. Assuming that the technical and political problems can be solved, the world schedule for DBS systems looks something like this:

- **1983**—Canada starts service from its Anik-C satellite. France and Germany launch two satellites for large-scale tests.
- **1984**—The Japanese launch their BS-II satellite and start their service.
- **1985**—First American DBS systems go into operation. The Franco-German system goes into full operation.
- **1986**—Radio Luxembourg begins broadcasting commercial television to France and Germany, thereby opening a whole new can of worms.
- **1985-1988**—Italy, Australia, and several other countries get their own DBS systems.
- **1989**—The Scandinavian countries launch a very high-powered satellite that can reach high into the sky over the Artic Circle.
- **1990s**—Nations all over the globe follow suit.

TECHNICAL PROBLEMS

The major technical problem with DBS is very simple. How do you build the receivers cheaply enough to be attractive to enough customers to make the system pay? Like many "simple" problems, this one turns out to be a real headache.

DBS is not aimed at the kind of people who buy home earth stations. That's a big market, but not a mass one in the way that color televisions are a mass-market commodity. A DBS system would need hundreds of thousands of potential viewers to be successful. DBS is expensive to set up, and you have to spread the cost over many customers to make it worthwhile.

It is a generally acknowledged fact in the consumer electronics industry that $1,000 and $500 are magic numbers. A product that costs more than $1,000 won't sell nearly as well as if it cost $1,000 or less. If you can get the price down under $500, your market multiplies several times over. Nearly every kind of consumer electronics product from color television sets to home computers has demonstrated this pattern. There is every reason to believe that DBS systems will, too.

Estimates of the cost of DBS equipment vary from about $160 to around $1,000, depending on which technology you believe in and what assumptions you make. No one who is predicting success for DBS figures it will cost more than about $1,000 to install.

A DBS system could be put into operation tomorrow. It would be economic suicide, though, because the receivers would cost so much. It would probably cost more to outfit your home with a DBS system than it would to install an earth station.

The problem is that DBS systems will use a band of frequencies nearly twice as high as the band used now for satellite television. Designing and building equipment which will work in that range is tricky even when you're not worried about building it to price. The receiver's job will be no more complex than the job done by an earth station receiver, but the frequency range makes things harder.

DBS receivers will need to use microwave technology more sophisticated than anything that has ever been used in a consumer product. It is on par with the technology used in modern military radars.

One thing is certain. DBS receivers won't be built the way Ku band equipment is built now. Today gear that operates on that band is practically hand-assembled and very carefully calibrated. To meet the price constraints of DBS, the receivers must be mass-produced with a minimum of calibration and alignment.

The companies working in the field have developed three basic approaches to meet this challenge. Some firms led by the Siemens company of Germany, are planning on using integrated circuits made with gallium arsenide (GaAs).

Gallium arsenide offers a number of advantages over silicon in building solid-state electronic devices. It is theoretically possible to build the critical parts of a DBS receiver as individual integrated circuits out of GaAs. Gallium arsenide devices are difficult to make and have been very expensive to date. A single GaAs transistor can cost as much as $300 compared with just pennies for a silicon transistor. Until 1982 no one had succeeded in making integrated circuits out of GaAs on a commercial basis.

Siemens changed that when it brought out a GaAs integrated circuit designed to be used in the low-frequency stage of a DBS receiver. The selling price for the integrated circuits was $100 each, but company officials claim they can cut that to $25 each if production reaches 100,000 a year.

Based on its success with the design, Siemens is looking forward to putting the entire front end of a DBS receiver on a single chip. Company officials say that the first DBS receivers will probably be hybrids that use a mixture of GaAs, silicon discrete components, and integrated circuits (ICs)

Other companies that are pursuing GaAs technology aren't so sanguine about a monolithic GaAs integrated circuit. They generally feel that DBS receivers will stay hybrids for some time to come.

The second approach is one pioneered by NHK. It uses conventional components, but it very cleverly combines the mixer and down converter into a single unit.

The mixer-down converter approach uses a diode coupled to the waveguide to convert the signal to a lower frequency. The result is an extremely simple and quite inexpensive design that is probably the furthest along of any of the proposed systems. NHK has licensed the process to at least 17 companies. Most are Japanese companies. Hundreds of receivers have already been built using this method. Its developer, Yoshihiro Konishi, claims that a receiver can be built for $160.

Konishi also claims that the inherent simplicity of the method, and the fact that it uses so few parts, means that it can amplify a signal with less noise than either of the competing methods. That may or may not be true. It is certainly disputed by companies pursuing other approaches. They claim that NHK has pushed the mixer-down converter approach about as far as it can go. Because the down converter-mixer method is simple, inexpensive, and has the backing of NHK, it is certainly a powerful contender.

The mixer-down converter's lead may not be as commanding as it appears. Of the 17 companies that have licensed it, nearly all are Japanese. Many of them are exploring other methods. Some companies are openly in favor of alternates such as microwave integrated circuits that use a front end built around field effect transistors (FETs).

Microwave integrated circuits (MICs) also use GaAs. Instead of trying to put everything on one chip, a combination of discrete components and integrated circuits is used to do the same job. One company that has been very active in the MIC approach is Plessey. Plessey has a design that uses a GaAs integrated circuit for a preamplifier, followed by an image rejection filter in GaAs, and a FET mixer followed by a more-or-less standard intermediate-frequency amplifier. This amplifier feeds the signal into a first local oscillator (built with surface acoustic wave technology), followed by a second intermediate frequency stage and another local oscillator that both use surface acoustic wave devices).

Sony is also pursuing the MIC approach. The company predicts that it will be able to sell hybrid receivers for about $500 each.

It's hard to say which of the competing technologies will win out in the marketplace. NHK's microwave-down converter system has taken a strong early lead. As production climbs into the hundreds of thousands, the monolithic GaAs ICs favored by Siemens become increasingly attractive. One thing that will have a strong influence on the outcome is the quality of reception that customers will demand. The NHK approach produces quality pictures, but not as good as the other methods are capable of supplying. If the mixer-down converter falters, the short-term winner will probably be the MIC system, with the monolithic IC approach replacing it later.

POLITICAL PROBLEMS WITH DBS IN AMERICA

The technical problems inherent in DBS are serious but manageable. The political problems are at least as serious and potentially a lot less manageable. DBS is a technology that offers something to offend powerful special interest groups nearly everywhere.

Domestically DBS has problems because it threatens to upset the system of television broadcasting that has developed in the United States over the last 40 years. The basis of that system is the local television station and the stations are none too happy about it.

Owning a television station in a major market in the United States isn't a license to print money, but it is a business that is virtually guaranteed a profit. Local stations with network affiliations make most of their money by selling commercial time during network programs. This is a very good deal from the stations' standpoint. That time is expensive, but advertisers are eager to buy. Some of the time is sold to local merchants. Much of it goes to the same national advertisers who also buy time on the network.

The networks know they can't exist without local stations. A network is its chain of local affiliates in a very real sense. FCC regulations limit the number of stations a network can own, so it has to have affiliates to offer the sort of national coverage that attracts advertisers. The network gives up a large chunk of commercial time to the affiliates and generally tries to see that the affiliates are kept happy.

Not all the benefits of this arrangement accrue to the networks and the affiliates. The FCC requires that local stations justify their licenses by trying to serve their communities. One result has been outstanding local news coverage by the stations. (Anyone who doubts that is invited to compare American coverage of local news with the coverage in other countries. Typically, there is no local coverage of cities other than the capital.)

DBS threatens to radically alter this practice. That makes it extremely unpopular with broadcasters.

A DBS station or series of stations would function as a national network with no local affiliates. All the advertising revenue would flow back to the DBS operator. DBS operators would avoid many of the community service burdens that local stations have to bear. A DBS system would be extremely profitable given enough subscribers.

Local stations see DBS as a potentially damaging competitor not just because of the revenues it could siphon off, but because it could make some basic changes in the way the television industry does business. For that reason the station owners trade group, the National Association of Broadcasters, has strongly opposed DBS systems in the United States. When the FCC announced it would permit interim licenses for DBS, the NAB issued a statement blasting the FCC as having "abandoned the interest of the public."

Just where the public interest lies is hard to figure. It is true that our present system of television broadcasting has given us local service that is unequaled anywhere in the world. It has also meant that large areas of the United States have either no television or poor reception of one or two channels. In recommending that the commission adopt interim regulations, the FCC staff said they did not believe DBS would do major harm to our present system of broadcasting.

The commissioners themselves were cautious. Commissioner Abbott Washburn said he didn't want to see the day come when "the programs all came from space."

"The fact that we have strong local stations is really the foundation of American broadcasting," he added.

There is another potential problem with DBS. People may be getting direct satellite broadcasts at the expense of another technology—high definition television.

American television produces a picture composed of 525 lines. The electron gun in a television set sweeps the face of the tube 525 times for each frame of the picture. The resolution of a conventional American television set is limited.

There is no intrinsic technical reason for that limit anymore, as anyone who has worked with a computer with a color monitor can tell you. It is the standard, though, and we all live with it. The Europeans use systems that have more lines to the frame, so they get better resolution. Pictures on European sets are sharper and clearer.

Ultrahigh resolution video systems have been developed as possible competitors to movies. Douglas Trumbull, the special effects wizard who helped create the movie 2001, has been working on a system for several years that he claims can give better results than film.

None of this seems to matter if you're looking at the world through a 19-inch portable. If you see a typical television image on a projection television screen, the lines are noticeable and very distracting.

NHK and Sony have developed a high resolution television system that scans the picture tube 1125 times per frame. That is more than twice the American standard. The result is an amazingly sharp and vivid picture even on a 19-inch portable. CBS has licensed the system in the United States and wants to start using it. NHK, Sony, and CBS all see high definition television as the system of the future. They say the improvement in picture quality will become important as big-screen television becomes more widespread.

The snag is bandwidth. High definition television requires a lot more bandwidth than regular television, and that pretty well excludes it from the part of the spectrum where conventional television resides. To get that bandwidth, CBS wants to move high definition television up into the Ku band—right where DBS wants to go.

This isn't such a coincidence. Both services need a lot of bandwidth. The Ku band is the next more or less unoccupied part of the microwave spectrum that can support a broadcast application effectively. CBS has used high definition television to argue against granting that part of the spectrum for DBS use. The fact that CBS is one of the companies which has applied for a DBS license is no more ironic than other things that go on in the satellite television business.

POLITICAL PROBLEMS WITH DBS IN EUROPE

Americans aren't the only ones who have political problems with DBS. In spite of their rush to establish systems, the Europeans have their own difficulties to sort out.

The problem in Europe is different. Most European television networks are owned by the government and operated as a kind of adjunct to the post office. There aren't any local stations in the American sense, so there aren't any station owners to worry about. There are neighbors, though.

There are many European countries, and the area of these countries is smaller than that of the United States.

Because no one has figured out how to put up an antenna that produces a beam exactly the shape of, say, France, DBS signals from one country will inevitably spill over into the surrounding countries. If your equipment is good enough, you will probably be able to pick up any European DBS channel from any country in Europe. This is not a prospect that thrills most European governments.

The most immediate concern for most European networks is the vest-pocket nation of Luxembourg. Unlike most of its neighbors, Luxembourg allows commercial advertising on its radio and television stations. Because of the competition that engenders and the revenue it produces, Luxembourg programming is a lot more lively than the official programming in nearby countries. (In fact, by American standards nearly all European television is in the serious to dull category.) Since the 1930s Radio Luxembourg has been broadcasting to most of Europe and giving its listeners what they want, not what their governments think is good for them.

Radio-Tele-Luxembourg (RTL), the country's broadcaster, now wants to go on DBS. It has announced plans to put a DBS system in operation by the mid-1980s to broadcast in French, German, and either English or Dutch. The cost is estimated at $350 million, but an RTL official calls the revenue potential "tremendous."

That's not the adjective that other European nations are using to describe RTL's plan. "Cultural imperialism" is more like it.

Advertising in Europe is governed by an often conflicting welter of rules. As the *Wall Street Journal* pointed out in an article describing the situation: "A Kellogg's Corn Flakes commercial aired in Britain would be banned in Holland because it boasts of 'extra vitamins'. It would be outlawed in Germany because 'best to you' sounds like a 'competitive claim'. It would be off the air in France because children can't endorse products there. And it would be nixed in Austria because children in that country can't even appear in commercials."

Beyond the problems with advertising, DBS television would further break down the already eroding barriers between the nations of Europe. Whether that's a good thing or not depends on your point of view. For some people it would help usher in an age of peace and cooperation on the Continent; it would speed the day when there would be a United States of Europe.

Others see it as opening the door to further homogenizing of Europe into a Common Market culture, one in which the distinctive heritage of the different nations would be lost. Many of the governments in Europe not surprisingly tend toward the latter view.

POLITICAL PROBLEMS WITH DBS IN THE THIRD WORLD

If the idea of direct broadcast satellite television makes Europeans uncomfortable and American station owners nervous, it is enough to give every dictator in the world nightmares. The last thing a dictator wants is to have his people exposed to a channel of information that he cannot control, and that is precisely what DBS is. Even if no one is broadcasting directly to the dictator's country, there is still enough overlap in most cases that a fair portion of his population could receive the signal.

There are more dictatorships, one-party states, juntas, and totalitarian governments on this planet than democracies. None of them are happy with DBS.

DBS is actually just part of a couple larger problems that have been occupying much of the attention of various third world countries for the last few years. One problem is something called the New World Information Order, and the other is the question of allocating scarce resources like the electromagnetic spectrum.

Most third world countries have felt for some time that they don't get a fair shake from the news media in the more developed nations. Part of this is the kind of pique that a dictator naturally feels when the world press reports that he has just had his political opponents taken out and shot. Part of it is a very legitimate concern on the part of all small nations that their news is filtered through organizations that don't understand them, don't care, and that frequently appear hostile to their aspirations.

The result has been a proposal, under the auspices of the United Nations, for what is euphemistically referred to as the New World Information Order (NWIO). If you remember your history, you'll recall that the last people to go around talking about a New World Order were the Nazis.

The NWIO is essentially an attempt to censor the flow of information into and out of every country on the globe at those nations' whims. A satellite beaming someone else's television into your country does not fit into this scheme of things.

DBS is cultural imperialism of the highest order to many third world nations. Many others are uncomfortable with the idea on general principles. They want to be able to develop their own cultures and not take them off the tube.

The other concern many small nations have is who is going to make what use of the electromagnetic spectrum. In the past the developed nations have more or less divided up the spectrum in whatever fashion they pleased. Now the third world nations are looking for a piece of the action. One of the most important pieces is the bands allocated for satellite communication.

While the number of available satellite channels is large, it is not infinite. Because all satellites have to sit in geosynchronous orbit on the equator, the nations of both hemispheres have to share the same area of space. Given the present technology, that area is limited. The developed countries are rapidly filling it with satellites for their own uses.

For a developed nation like the United States, communications satellites are very handy things to have. They can be vital for developing nations.

Having a transponder on a Comsat means you can link a nation together without having to string land lines. It means being able to reach out into remote villages to teach people reading and the rudiments of sanitation. It means reliable communications and remote diagnosis for clinics run by paramedics.

A DBS system is particularly well suited to rural education projects. The antennas are small enough that the entire set can be packed into the back of a jeep and taken where it is needed. The DBS system is cheap enough that even a poor nation can afford to purchase the receiving equipment.

If you're from a poor country, it is awfully hard to justify using this new technology to add more television channels in the United States when 30 percent of the children in your own country under a year old die because people don't know how to dig a proper well. You can also argue that without the mass market represented by the developed countries, DBS systems will never be developed. You can claim that without an economic return, DBS satellites will never be launched.

You can also claim that the real interest of most third world governments is in getting involved with satellite communications at the expense of the nations that developed satellite communications in the first place. You'd probably be at least partially right. The electromagnetic spectrum should be shared according to some scheme.

A major portion of the sharing is expected to take place in the summer of 1983 at the Western Regional Administrative Radio Conference in Buenos Aires. These meetings are held every so often to allocate frequencies for various uses. One of the uses being proposed this time will be DBS. No one is sure what is going to happen, but the third world nations of the Western Hemisphere have indicated they intend to press strongly for more frequencies for their use. To add interest to the occasion, the host country, Argentina, will probably still be nursing a grudge against Britain and the United States over the Falkland Islands War.

DBS AND RUSSIA

Discussions of DBS television have been going on in the United Nations for the last 10 or 11 years, as part of the debate over the peaceful uses of outer space. The major bone of contention has been the rules under which one country would be able to broadcast via satellite to another country.

The third world nations and the Soviet bloc are demanding a provision that would prohibit such broadcasting without prior approval of the country receiving the broadcast. The United States has been resisting on the grounds that it would inhibit the free flow of information.

The concept of free flow of information cuts both ways, and it is entirely possible that the United States could find the edge very sharp. Under the rules the United States is proposing, there is nothing to prevent the Soviet Union from putting up a DBS satellite to broadcast to the United States. This possibility was raised in recent congressional hearings on the FCC.

If DBS becomes popular in this country, the Soviets may find this to be an irresistable opportunity. It would give them a propaganda channel into millions of American homes unfettered by any kind of government interference. It would very neatly hang us with our own rope—something sure to score points with third world countries.

Just how effective it would be in influencing public opinion is something else again. Russian broadcasting tends to be dull, as does much of Russian propaganda. It's hard to imagine many Americans tuning in to watch a tour of a Soviet tractor factory.

If DBS technology was combined with a sophisticated approach to information, it might become at least an irritant for our government. This is especially true because the United States would not be able to turn the trick against them. The Soviet people are extremely unlikely to be given access to DBS technology.

PIRATE DBS

A bigger irritant would be a free enterprise satellite in the tradition of the pirate radio stations of Europe or the Mexican border stations. European radio is usually state-controlled and dull, and there was a tradition of free enterprise radio broadcasting commercial programming someplace where the affected government couldn't reach it. The pirates were particularly active in the 1950s and 1960s. Some operated from ships moored off the coast of Europe, and some operated from countries with more lax attitudes.

During the 1960s the pirates were very popular because they were willing to broadcast rock music that was banned in their target countries. That got them a big youth audience and a lot of advertising. The stations were finally closed down because of the growing unity of Europe under the Common Market. The increased cooperation in Europe destroyed the sanctuaries they needed to operate.

A similar situation exists (or existed) in Mexico. Although American restrictions on broadcasting were much more lenient, there are still things you can't say or do on the air here. This is especially true with regard to some kinds of advertising. The quack doctors, shady preachers, and the rest moved to Mexico. The result was a chain of stations in border towns that broadcast directional signals at high power into the United States. At night you could hear them from Arizona to Alabama pitching their nostrums, exhorting contributions from the faithful, and playing popular music of a particularly kitschy sort. The broadcast operations proved to be extremely profitable in both Europe and the United States. They enabled advertisers to reach an audience they couldn't otherwise touch, and they were willing to pay heavily for the privilege.

Imagine a pirate DBS satellite bringing you television you normally can't get here and being supported by advertisers who you wouldn't see otherwise. It could happen very easily. In spite of the fairly loose restrictions that apply to American television, there are still some things that aren't going to go out over the air as long as the FCC has anything to say about them. Pornography is the obvious example. Whether you like it or not, there is an enormous market for porn in this country. Some people suggest that X-rated video cassettes have made the video recorder industry. That may be an exaggeration, but it is certainly true that the widespread use of video recorders has been a boon to the pornography industry in this country. Many "adult" video cassettes are sold.

Although some cable channels show R-rated movies as extra-price features, there is no sign that the kind of X-rated material available on cassette will ever make it onto the cables. This possibility was discussed in the 1982 congressional oversight hearings before Congress. Both the FCC commissioners and the congressmen were adamant that cable must be kept clean. There is nothing to stop another nation from putting an X-rated movie channel on a DBS satellite or other programming most people would consider undesirable.

The third world countries in the Western Hemisphere are almost sure to get some slots for satellites at the Western Regional Administrative Radio Conference in Buenos Aires in 1983. A Central American or Caribbean nation could contract with the Europeans to launch a bird with, say, 12 video channels available. Most of those channels would be used for communications, education programming, and other squeaky-clean purposes, possibly on a regional basis. Two or three channels would be set aside for commercial use supported by advertising. These channels would be beamed at the United States and would carry the most terrible selection of programming imaginable. A selection of X-rated movies could be played right off the cassettes, interspersed with ads for cancer cures, secret ways to make a million dollars, Central American lottery tickets, and medicines to increase sexual potency.

The United States government would howl, but the third world countries would stand firm. They could point out that the United States was so adamant about freedom of the skies. Besides, the commercial channels would be defraying the cost of the satellite and providing many vital services to the third world nations. Most of their people are too poor to own DBS setups, so they wouldn't have to worry about corrupting their citizens.

The best that the United States could hope to do would be to keep some of the more violently objectionable material off the air. The pornography would continue to flow.

We don't know that anyone is considering pirate DBS or that it would develop in that way. But it is a distinct possibility given the potential of the medium, and it is something that should be kept in mind as the DBS debate develops. DBS-TV may give you a much wider window on the world than you now envision or possibly even want.

No one knows for sure what the fate of DBS will be. Our guess is that the technical problems will be solved and the political difficulties overcome. By the end of this decade, DBS systems will be as common as cable television is today.

Chapter 11

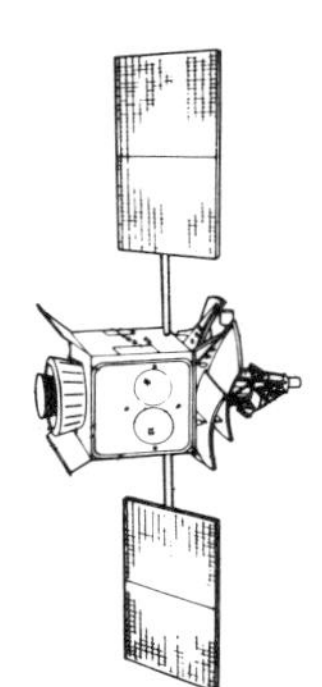

The Legalities

By now we hope you see that home satellite television is a rapidly burgeoning field with much promise (Fig. 11-1). You're probably wondering just how legal it is.

Unlike regular television or radio, satellite signals are not being broadcast specifically for home reception. Isn't it against the law to pick up all that programming? Couldn't the owner of a home earth station be sued or arrested for what he's doing?

Those are very good questions. We can't answer them, and no one else can either.

At this time home satellite television is not against the law. No one can arrest you for having your own earth station. There are some serious unresolved questions, though, about using an earth station to receive some of the signals. These questions have not yet been answered in the courts, and Congress has declined to act on them. There is no law on these points. That does not mean there aren't some strongly held opinions.

Before we go any further, let's make a couple things absolutely clear. We are not lawyers, and we're not attempting to pass ourselves off as experts in an area that is legally very complex. In this chapter we are attempting to sum-

Fig. 11-1. RCA's Satcom IV is prepared for launch at Cape Kennedy, Florida (courtesy RCA Government Systems Divison).

marize the debate going on about home satellite television and offer our predictions as to the probable course of events. Take it for what it is worth, but don't treat it as gospel.

We cannot and do not condone taking someone's property without recompense. Theft is theft whether it involves something as tangible as money or as intangible as information. If you're not legally entitled to it, you shouldn't take it.

This is important because the debate that's going on right now does not involve home satellite television. The

FCC has ruled that home earth stations are legal. The problem is with the unauthorized reception of signals by home earth stations, specifically signals intended for cable or other pay TV outlets.

There are many broadcasters who don't care if home earth stations get their signals. Some are religious outlets that want to spread their message as widely as possible. Others are superstations that make their money from advertising. The more viewers a superstation has, the more it can charge its advertisers. Some advertisers are very interested in reaching viewers with home earth stations.

There are usually some of those "don't-care" broadcasters on every satellite in the sky. This complicates matters considerably for the ones who do care.

PAY TV SERVICES' CONCERN

The pay TV services oppose home earth station owners picking up their signals. They believe it is theft. Someone with a home earth station can get their signal without paying for it and that, the services feel, is illegal.

The pay TV services' concern seems a little exaggerated at first glance. One hundred thousand earth station owners can't make much of a dent in a video viewing audience that numbers in the tens of millions, especially when many of them are located in rural areas that will never have pay TV.

There are several reasons for the pay TV services' concern. At the rate the home earth station business is growing, there will be millions of them in the next few years. Also, not all the home earth stations are located in areas without pay TV. Each home earth station that can pick up pay services cuts into the profits of the local operators. Some of those earth stations are master antennas that serve an entire apartment complex or hotel. In that case the loss of potential revenue from a single earth station can be significant.

Finally, most material that the pay services use is covered by copyrights owned by someone else. The services feel they have a duty to protect those copyrights. If they don't, they argue, movie studios and other copyright holders will be reluctant to deal with them or will charge them much higher prices for the right to use their property.

The future expansion of pay TV services will take place increasingly in small and medium-size cities. Most big cities now have one or more pay TV channels, and the ones that don't are getting them. There are still millions of homes in those areas that haven't had service exended to them yet. The new franchises are going to be in the smaller, less densely populated urban areas.

There is a direct relationship between the cost-per-customer of providing cable television service and the population density. That is one of the reasons that cable television first became popular in Manhattan, the most densely populated major urban area in the United States. When you start extending service to less densely populated areas, your costs go up. If there is a significant number of home earth stations in town, representing people who (presumably) won't be interested in signing up for cable, the cost of setting up a cable system can be increased significantly by delaying the break-even point. Home earth stations might even make cable economically unattractive and prevent any sort of service.

This is a matter of concern for the companies putting pay TV signals up on satellites for two reasons. First, they feel that their economic health is tied to that of the pay TV operators. Second, many of them are owned by companies that also own pay TV systems. Home Box Office is owned by Time Inc., which also owns American Cable Television, the largest cable operator in the country.

FREEDOM OF THE AIRWAVES AND HOME EARTH STATIONS

All this takes place against a rather peculiar legal background caused partly by an odd provision of the FCC regulations. There is no law in this country against receiving any kind of electromagnetic signal.

It is perfectly legal to listen in on the president of the United States when he uses the radiotelephone on Air Force One to communicate with the White House. There are radio amateurs who listen in as a hobby. Some radio amateurs have some fascinating tapes of conversations between high government officials. In most countries that sort of action would get you a long jail term or shot for espionage, but in the United States it is legal.

There have been many attempts in recent years to pass laws prohibiting ownership of police-band receivers on the grounds that they can aid criminals. Generally they haven't fared well in the courts. The FCC has specifically said that it is not against the law to own equipment which can receive on emergency frequencies. There is nothing to prevent a private citizen from listening to police calls.

This has led to the "freedom of the airwaves" argument. The basic idea is that it is legal to tap any signal on the air. If you want to keep something private, you either shouldn't broadcast the information or you should scramble it so unauthorized listeners can't understand it.

This argument carries a lot of force, especially with people who are familiar with ham radio. Most hams hold with it, at least in part. Certainly it has a strong emotional appeal in "the land of the free." It is widely accepted that the airwaves are a natural resource, and there's something repugnant about the idea that we can't do what we want with our own receivers.

The situation isn't nearly that simple. While the FCC says it is legal to receive a signal, it places limits on what you can do with the signal you receive. Even though the freedom of the airwaves argument may be emotionally attractive, it generally hasn't fared well in court.

As far as the FCC is concerned, the limits on the use of a received signal are embodied in section 605 of the Communications Act of 1934 (47 USC section 605). According to the FCC, section 605 makes it unlawful:

—"For a person not authorized by the sender to intercept radio communications and divulge or publish the existence, contents, purport, or meaning thereof to any person; or,

—"For a person not entitled thereto to receive radio communications and *use such communication or any part thereof for his own benefit* or for the benefit of another who is not entitled thereto."

That section has a long history, going back to the first radio act of 1912. There are also several cases where the courts have interpreted it.

That would seem to settle the issue, but it doesn't. Instead, things are so unclear that the cable companies and others concerned with pay TV have been going to Congress every session for the last few years and asking for a new law.

The question is how section 605 should be interpreted regarding home earth stations. What constitutes unlawful use of the signal? Is unauthorized reception for home entertainment unlawful? Is it unlawful to pipe the signal into an apartment complex? What about putting it on big-screen television in a bar?

There are only a handful of cases to date dealing with unauthorized reception of microwave signals from local pay TV (MDS) broadcasters and even fewer dealing with private earth stations. As far as we know, none of the earth station cases has involved a home earth station being used by a single household. None of them have been finally decided.

One problem is that "traditionally," private use of a signal for noncommercial purposes has been held not to be a violation of section 605. That is why thousands of peope make a hobby out of listening to police and fire monitors. There are even some uses of unauthorized receptions for business purposes that the FCC doesn't choose to pursue. Nearly every newspaper and television station in the United States has one or more scanners tuned to emergency frequencies, and the FCC has not objected.

The exception to the exception is when the signal is privately used for the same purpose for which it was originally intended. Listening to the police bands for entertainment isn't putting the signal to the use for which it was originally intended. Tuning in on a pay TV company's microwave broadcast to watch a movie is.

You can argue that unauthorized reception of a satellite signal by a home earth station is putting the signal to the use for which it was originally intended. You can also argue that the satellite signal is a common carrier, and using it for home entertainment is not putting it to its intended use. (Technically Comsat users are common carriers, not broadcasters.)

The situation with the "freedom of the airwaves" argument is much clearer. There have been several cases where it has been raised, mostly involving local microwave pay TV broadcasts, and it has lost.

In a 1981 case a federal district court in New York held that a pay TV company does not have to scramble its signal to establish that it did not intend for anyone to be able to use its signal. In part the court's opinion said:

> "HBO's failure to restrict the reception of its transmissions by scrambling its signal cannot be given the significance attached to it by the defendants. There is no indication in the statutory language, or in the legislative or regulatory history, to suggest that Congress meant to protect only those persons who take affirmative steps to keep transmissions confidential."

(Home Box Office Inc. v. *Advanced Consumer Technology, et. al.,* (DCSDNY), 81 Civ. 559).

There are arguments on both sides of the question of legality. You can argue all you want. Until a definitive home earth station case makes its way to the Supreme Court, though, no one will really know.

It is suggestive of the problem's complexity that to date the pay TV companies and others opposed to home earth stations have not gone after the manufacturers of satellite receiving equipment and their customers the way they have attacked the makers and sellers of local microwave receivers. Most earth station cases that have been filed were instituted by boxing promoters attempting to stop bars from showing major fights. The issue of commercial use seems clearer there than in the case of an earth station for home use.

Some manufacturers of earth station equipment and representatives of their trade organizations suggest the reason for the reluctance is the programmers' fear of what would happen if they lost. Whatever the reason, no one is suing owners of home earth stations—at least not as this is being written.

If the pay TV services choose not to act under the

present law, they have two alternatives. They can try to get a new law, or they can seek a compromise with the earth station forces. In fact, they are seeking a new law.

HR 4727: CHANGING THE LAW

In October 1981 Representative Henry A. Waxman of California introduced HR 4727, "A Bill to Amend the Communications Act of 1934 to Establish Penalties for Violations of Section 605 of the Act," as it says in the title.

The bill and the hearings that have been held are interesting. In the opinion of the proponents of home earth stations, the bill represents an attempt to kill the industry by levying huge fines on the manufacturers and users of home earth station equipment.

We say "in the opinion" because nowhere does the bill specifically refer to home earth stations. That is one of the things that makes it so interesting.

HR 4727 established civil and criminal penalties for persons violating section 605 of the Communications Act. It gives the aggrieved party the right to sue and provides for heavy damages under a very liberal standard of proof.

The plaintiff is allowed to recover all the defendant's profits if the court finds him guilty. Also, the plaintiff gets to establish the defendant's profits simply by making a showing of the defendant's income. The defendant must try to prove his expenses.

The bill also provides that the court may award additional damages of up to $50,000 if the defendant willfully and for purposes of commercial advantage and financial gain violated section 605. There are criminal penalties, including up to a year in prison.

The bill does not attempt to define what constitutes a violation of section 605. It makes no mention of home earth stations, local microwave receivers, or any other kind of technology. It certainly does not attempt to outlaw home earth stations directly.

Representative Waxman repeatedly stressed that the purpose of the bill was not to make home earth stations illegal. Both he and the witnesses for the cable companies said that the targets of the bill were the video pirates who were selling receivers that could tune in the local microwave distribution systems (MDS) used to transmit pay TV in some areas. Home earth stations were not the target, they said.

That was not the way the earth station industry interpreted the legislation. While the bill did not specifically include home earth stations, it did not specifically exclude them either. Representative Waxman and the witnesses representing the cable companies said they did not think it would be a good idea to change the language of the bill to exclude home earth stations.

According to the home earth station industry, the bill could effectively kill the manufacture of home earth stations in this country. Earth stations aren't mentioned, but the civil provisions are an open invitation for the cable operators and movie companies to bury the manufacturers and sellers of earth station equipment under a barrage of lawsuits. At the present time, federal law does not clearly establish a private individual's right to file a civil action under section 605.

This threat of liability doesn't just apply to home earth stations. Because section 605 applies to the unauthorized reception of any type of signals, the penalty provisions can be applied to manufacturers, sellers, and users of ham radio equipment, police band scanners, and other similar devices.

WHY THE HR 4727 APPROACH?

At first glance, HR 4727 seems to be doing things the hard way. If the problem is that home earth stations violate section 605 when they receive some signals, then why not just make it illegal to receive those signals? Or outlaw earth stations altogether?

Part of the reason is that it presents an almost insurmountable burden of proof. The second approach would never clear Congress.

Unlike the pirate MDS receivers, home earth stations can be used to receive a wide range of legitimate programming. Many broadcasters with signals on the satellites don't care who listens in and will willingly give permission for home earth station owners to receive their signals. Because there is at least one such user on every satellite, a home earth station could be pointed at any satellite in the sky and still be getting a legal signal. The only way to establish that a signal was being stolen would be to sneak up and peek in the window.

A closely related difficulty is that only the earth station operator could be prosecuted under that approach. With HR 4727, the fact that one earth station had been used to illegally receive a signal could make the manufacturer and seller of the equipment liable for stiff penalties. This is very much preferable from the standpoint of the cable company, because it allows the company to go after bigger game.

As for just outlawing earth stations, that doesn't appear to be politically possible today. There are too many congressmen from largely rural districts whose constituents have earth stations as their only source of television.

When the House Subcommittee on Telecommunications, Consumer Protection, and Finance held hearings on HR 4727 in November of 1981, the leadoff witness was Representative Charles Rose of North Carolina. Representative Rose was adamantly opposed to the Waxman bill and, among other things, had this to say:

"My colleague's bill threatens to make criminals out of those who use earth stations to watch subscription services in the privacy of their homes. I believe every American should be entitled to watch programming that comes into his backyard free of charge, really. If you put the signal in my backyard, I think I ought to have the right to watch it. If you do not want me to watch it, do not put it in my backyard.

Not even all the members of the subcommittee agreed with the bill, as Representative W. J. (Billy) Tauzin made clear in his examination of several witnesses. Shortly, we'll get to what Rep. Tauzin had to say, because it concerns one of the most ironic parts of the entire legal situation. For now, the point is that legislation affecting the rights of home earth station users is not popular in Congress.

HR 4727 went nowhere in 1982. Similar legislation will probably be introduced in the next session of Congress and be reintroduced in succeeding sessions. Unless provision is made to protect home earth station users, it is given little chance of passage. With each passing year, there are more earth stations and more people with an interest in protecting their right to watch satellite television.

Undoubtedly the cable and movie industries will eventually get legislation giving them more protection from video pirates. It will probably have to be drawn so that normal use of earth stations is excluded.

THE TECHNOLOGICAL FIX

Assuming that the programmers can't get the legal protection they want, what is left for them to do? The obvious answer is for them to scramble their signals. That is what some of them are doing or will be doing in the near future.

A television signal is transmitted according to a complex set of standards for frequency, pulse shapes, placement of audio, and other things. If you distort those things or mix them up, a conventional television receiver is hopelessly lost.

This principle is already used by many cable companies to control access to their signals. One popular device is to interfere with the blanking pulse that appears between frames of a television picture. The picture cannot be stabilized without a pulse of the proper amplitude and shape. If you combine that with a circuit which moves the audio signal from where you expect to find it, you have fairly effectively prevented someone without a decoder from watching your channel.

Many electronics magazines carry ads for do-it-yourself decoders that will allow you to unscramble the signal without paying the cable company. Anything that is encoded can be decoded, given enough time and equipment. The very thing that makes a television image so easy to scramble makes it easy to figure out how it is being done—that is, the high degree of signal redundancy. The person setting out to crack the scrambled signal knows there will be a blanking pulse between each frame, that each frame will consist of a certain number of lines interlaced in two half-frames, and so on. The only thing he won't know is the variations in voltages that determine the exact picture being transmitted and the sounds on the audio track.

Satellite television offers a major advantage over home television when it comes to scrambling the signal. The cable companies have relatively few authorized receivers—only one per cable service area. That means they can afford to build a much more complex and expensive scrambling system than can be economically installed on an individual television receiver.

Remember anything encoded can be decoded. If you can spend a lot more money to encode your message than your opponent can spend to crack it, you are reasonably secure.

With satellite television, some of the proposed encryption schemes are so complicated that they will be effectively unbreakable by someone with a home earth station. (Theoretically you could tape the scrambled signal and "massage" it at your leisure. But that's a lot of work.)

One scheme we have seen proposed involves changing the encryption key every few minutes by transmitting the new key over a leased telephone line to each legitimate earth station. Another involves using specially designed semiconductor chip sets to scramble the signal by inserting a variable length delay before the beginning of each scan line. The length of the delay would be governed by a pseudo random number sequence built into the chip set.

One of the first systems to go into use is much simpler. The company has rented transponders on two satellites and will put the audio on one bird and the video on the other. Receiving the signal will require either a spherical antenna with two LNA-receiver sets or two completely separate installations if the satellites are more than about 15 degrees apart.

This simple system has the advantage of being cheap and reasonably secure, because few earth stations can pick up two satellites at once. (If you and your neighbor both have earth stations, one person can pick up the audio, the other can go for the video, and you can share the signal.)

There is no theoretical reason why all programmers

couldn't scramble their signals. The only ones who seem to be interested in doing so are the premium service providers. Even they don't like it.

The reason is that scrambling increases the cost of receiving the signal to the cable company. Because cable programming services are highly competitive, the added cost of decoding the signal could cut the number of cable systems subscribing to a program service. (Some industry sources have claimed it could cost a million dollars or more per cable system to set up a scrambled system.) That is also why putting the audio and video on different satellites is so attractive as a scrambling system. It is cheap for the cable companies to "decode." For the companies that are genuinely concerned about pirates, scrambling seems to offer a practical method of protection.

ANTITRUST, SPACE, AND SCRAMBLING

So far we've been talking about a home earth station receiving signals from a programming service that has not authorized reception. Suppose the earth station owner wants to do things legally. Perhaps he writes the programmer and asks to sign up for the service at the regular cable rate.

The earth station owner will usually receive a letter by return mail telling him the service is not available. The tone of the letter will vary from friendly regret to threats of dire legal consequences if he doesn't stop watching that channel. The home earth station owner usually cannot pay for premium program services even if he wants to do so.

It's hard to determine why this curious state exists. Part of it has to do with the terms of the contracts between the programmers and the holders of copyright on the material they use (such as movie studios). Those contracts specify that the programming is being provided for sale to cable systems. Selling the service to home earth stations would presumably violate that copyright.

This is probably accurate in the short term. There is no reason why those contracts have to exclude home earth stations. Neither the premium programmers or the copyright holders are particularly interested in making provisions for the obvious fact that the home earth station owners are going to be tapping in. This is particularly odd in the case of earth station owners who are willing to pay for the service.

The oddity has been repeatedly noted, particularly by the proponents of home earth stations. In questioning witnesses during the congressional hearings on HR 4727, Representative Tauzin put it this way:

"If you are going to penalize a backyard viewer, a backyard satellite viewer, and you are going to say he is going to be subject to a fine for having pirated that signal under this legislation and not tell him at the same time, look, if you want to be honest and pay for the signal, you are entitled to it, are you not meeting only halfway with the American public?"

The other argument that the programmers make is that they are not set up to collect individual payments from thousands of subscribers all over the country. The bookkeeping chores, they say, would be enormous.

One group that has proposed an answer to this is SPACE, the *Society for Private and Commercial Earth Stations.* SPACE is a trade group for manufacturers of earth station equipment. It is a voice in Washington for earth station owners. (The address is 1920 N Street NW, Suite 510, Washington, D.C. 20036.) SPACE has essentially proposed setting up a royalty fund similar to the ones handled by BMI and ASCAP in the music industry. Royalties would be collected from earth station owners either as a one-time fee included in the price of the equipment or on a regular basis. The money would be divided among the holders of copyrights on the material shown on subscription channels.

The programmers are generally not at all interested. Most of the major services have rejected the proposal. The services contend that the government would end up regulating copyright payments from home earth station users to the copyright owners. Also, they are under no obligation to provide service just because someone wants it.

The first contention may or may not be true; certainly no one denied it very strongly in the hearings on HR 4727. The second contention raises some antitrust issues that an attorney would describe as "very interesting."

The antitrust issue comes into play in two ways. First, a refusal to deal is usually considered to be anticompetitive. The programmers will not sell their service to some people who wish to buy it. Is that anticompetitive within the meaning of the antitrust laws?

No one knows. Antitrust is a very complex area of the law and without court cases to fall back on, it is very difficult to say. David Aylward, the chief counsel to the House Telecommunications Subcommittee, has his doubts.

"Unless a company has real market power, a refusal to deal is not a violation of the antitrust law," Aylward said in the September 1982 issue of *SatGuide* magazine. "It's not clear to me at this point that, for instance, Warner Amex has a monopoly of anything."

"The fact that program suppliers are linked up with cable operators is only a problem if they monopolize the market. If they don't, you've got other places they can go," Aylward added.

Aylward was referring to the second potential antitrust problem the subscription services face. Many services are owned by companies who also own cable television operations.

For example, HBO is owned by Time Inc., the company that owns American Cable Television. The Movie Channel has common ownership with Warner-Amex. Showtime is tied to Viacom Cable TV and TPT/Westinghouse.

In theory cable companies and home earth stations are in competition. SPACE and others in the home earth station industry maintain this is the real reason why the programmers don't want to sell services to home earth stations.

Whether that belief is true or not, and whether it has any antitrust implications if it is true, remain to be seen. In the next couple years, the problem of home earth stations and subscription television has to be resolved.

That resolution may come fairly quickly. When HBO and some other programmers announced they were going to start encoding their signals, SPACE announced it would file an antitrust suit against them if they do. No one knows what the outcome would be, but the potential damage it could cause the subscription television services and their affiliated program providers is huge. Such a suit could serve as a powerful bargaining tool to come to some settlement of the issue.

OUR BEST GUESS

No one knows where the legal battles over home earth stations will end. At this point, however, a couple things seem fairly well established.

First, home earth stations are not illegal. You can watch most programming available from satellites legally. There are legal questions about tuning in most of the premium services, and the law in that area needs further refinement.

The programmers and copyright holders will likely come to some type of agreement with the earth station owners, probably along the lines suggested by SPACE. In the future you might have to pay an annual fee to use a home earth station, or an additional fee might be tacked on to the cost of an earth station. Some of the premium services may scramble their signals and sell or rent decoders to home earth station owners.

There are many alternative scenarios. For example, a bill like HR 4727 might become law. The premium services might go ahead and scramble, cutting them off from home earth station owners. One unlikely outcome is that you'll never be barred from putting up an antenna in your backyard to receive some satellite television.

Chapter 12

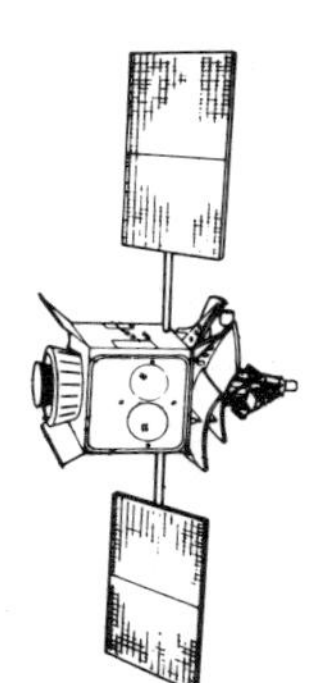

Social Implications

The use of satellites to broadcast information to the world's population has implications that reach far beyond the entertainment offerings dominating satellite television broadcasts in North America. There is little doubt that satellite communications are extremely cost effective and versatile regardless of the application. Modern microelectronics technology will continue to drive prices downward.

SATELLITE TELEVISION AND THE THIRD WORLD

The flexibility and lower prices will have a tremendous impact on the way that satellite communications help to shape our world—for better or worse. More than 100 countries are signatories to the Intelsat agreement. About 10 nations now have national satellite communications systems, and many more are now on the drawing boards. The use of satellites is no longer limited to rich, industrial nations. Some African nations are planning a satellite system, along with a group of South American nations, India, the Phillippines, Indonesia, and some Nordic countries (Fig. 12-1).

What would a third world nation, faced with illiteracy and a poor standard of living, want with satellite television systems? A satellite television system would entertain those

Fig. 12-1. Villagers in Kerelli, India, about 500 kilometers southeast of Bombay, received television for the first time in 1975 under a test program run by the Indian Space Research Organization using the American ATS-6 satellite. It was the forerunner of a program that will use satellite technology to bring change to villages all over the world (courtesy NASA).

who can afford it, but it would help address some problems of education, medicine, agriculture, weather forecasting, and mass communications.

Many third world nations have a few relatively modern cities (where the rich and the governing classes live), surrounded by vast distances of sparsely populated agricultural lands, where generally illiterate people try to eke out a living at the mercy of climate, poor soils, and little or no help from those in the cities. In many parts of the world there are people who are literally tied to a lifetime of poverty, malnutrition, and misery.

Imagine that you are in one of these villages and are fully dependent on agriculture. There is no running water or electricity.

Imagine that the government has managed to put up a satellite with six television transponders. It uses one of the transponders for telephone service, one for military communications, and the rest for entertainment (supplying the broadcast outlets in the cities with programming).

Suppose the Ministry of Education gets control of one of these transponders for whatever purposes it deems worthwhile. What could be done?

Let's return to the village. One day a government truck pulls up and begins unloading boxes of equipment. A few days later a 6-meter dish has been erected, along with the associated electronics and a television monitor. The equipment is highly simplified and capable of only tuning and receiving the one transponder's signals. The whole affair is powered by an array of solar panels, meaning that it will operate during the daylight hours only, and perhaps a few hours into the evening if battery backup is provided.

In a short time the countryside is dotted with such setups, each a few day's walk apart. What does the government provide its people? Major league baseball from the United States? How about a selection of the latest horror films to come out of Hollywood? Better yet, why not a selection of the 10 best political speeches from the ruling government?

Weather Forecasting

Let's say that the government began by providing the villagers with very simplified weather reports and advice on planting. With careful planning and equally careful scripting, the weather forecasts could become a powerful tool in helping the farmers become less dependent on nature's whims.

Along with the forecasts, the government could begin to teach modern agricultural skills to the people, helping them to improve the yields and quality of their crops. In most developing nations the land is not used to its fullest extent. The productivity of the farmers is reduced. In many parts of the world the use of pesticides is unheard of, the rotation of crops is an art not yet learned, and the construction of irrigation systems is quite primitive. With the satellites being used to deliver information, the farmers can begin to change their view of agriculture from that of a way of life to that of a profitable business. The benefits to the nation are obvious as malnutrition decreases and overseas food purchases begin to decrease, helping to reduce the national debt.

Education

Many developing nations managed to completely miss the Industrial Revolution and are locked into an agricultural economy. The gap between the rich ruling classes continues to increase, threatening national stability in many parts of the world. Even the oil-producing nations—with some of the highest per capita incomes in the world—are still plagued with problems of extensive poverty and malnutrition. The problem is often one of education.

Imagine that the transponder serving the rural population broadcasts daily programming that teaches the population how to read, write, and count. The process would be slow and expensive, but in time the illiteracy rate would begin to drop.

Imagine that one hour a day is spent teaching modern hygiene and medicine to the village healer. Knowledge built on a foundation of superstition and religious hocus-pocus is replaced with a foundation in proven sanitation and healing

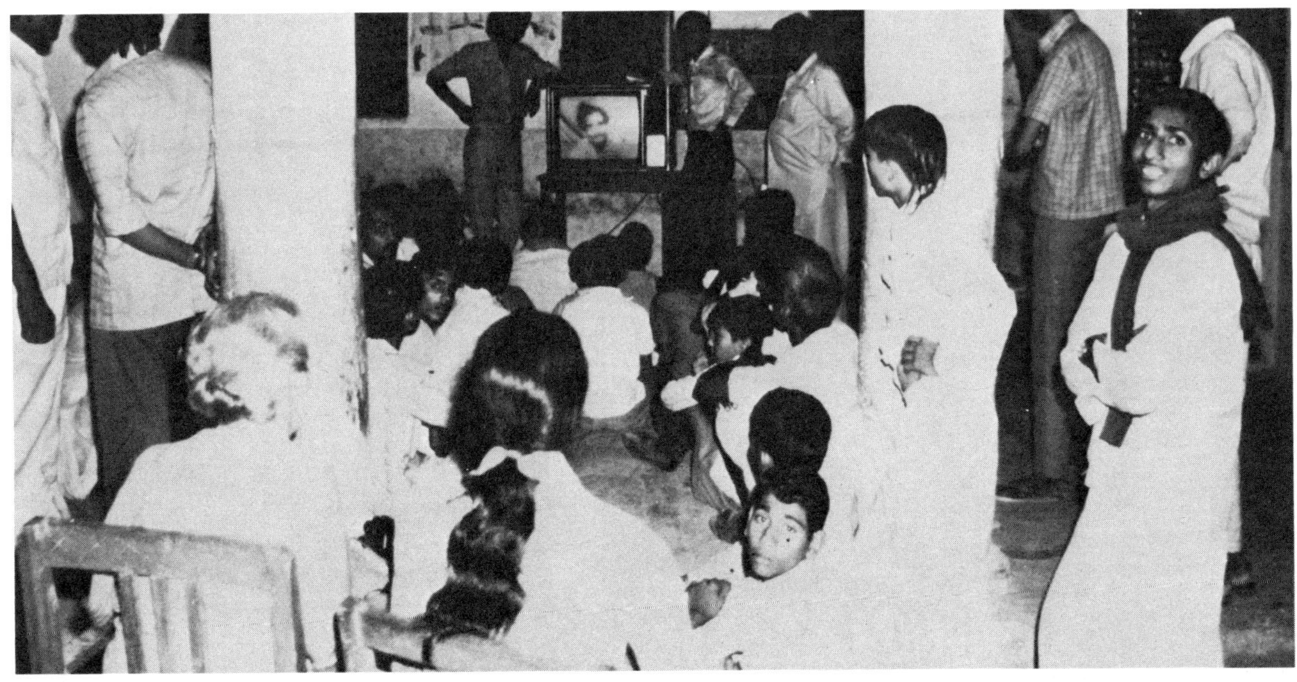

Fig. 12-2. The Indian government is one of many third world countries experimenting with using satellite television as a medium of education for its people (courtesy NASA).

techniques. The satellite could be used to spread health information in the event of an epidemic or plague, which has the potential of killing many people.

There can always be abuses, especially if some local despot decides to use the communications technology to guarantee his or her political future. Rebels will surely want to commandeer the broadcast center for a nationwide satellite television system. While abuses are certain to happen, there is little doubt that such information and education centers will prove invaluable in helping the emergence of many third world areas (Fig. 12-2).

SATELLITE TELEVISION AND THE UNITED STATES

It's nice to talk about how satellite television can be used in other parts of the world, but there are many more uses close to home, especially since the satellites are already on-station above the equator in orbits that give a broadcast footprint over most of North America. When people discuss and write about the United States, it is easy to think only about New York, Boston, Los Angeles, Phoenix, Chicago, and other large metropolitan areas.

Aid to Rural and Impoverished People

Many people live in the rural areas of the country, though. Those people have needs that satellite television can address. Using satellites to provide in-depth agricultural weather forecasts could prove to be priceless in the grain belt of the United States. The emergence of cultural and fine arts programming will enrich the lives of those people in communities not large enough to afford or draw major cultural attractions. The satellites could be used to provide the nation's farmers with up-to-date agricultural information.

There is a segment of the American population that lives in near isolation and often in near poverty. The large Indian population could make excellent use of satellite television, not only as a tool to help them assimilate into the white man's culture, but also as a way to solidify the native heritage.

Other areas of the country that are stricken by poverty

and illiteracy, such as Appalachia, could also benefit from the same treatment. Remote areas of Alaska could also benefit from education via satellite.

Political Uses

This book was written just prior to national elections. The domestic political uses for satellite television are staggering. The people in Washington will likely make sure that satellite television rules provide for plenty of political access.

Smart campaign managers could resort to having "documentaries" developed that are little more than blatant campaign pitches, hoping that some satellite network would broadcast them up to fill the dead space between features. Special interest groups could also try the same tactic, developing "shorts" that espouse their particular viewpoint. As you can see, there are some potentially negative implications for satellite television.

Civil Defense

Our satellite television system could have some potential as a civil defense tool. In the event of an outbreak of hostilities, the civil defense signals could be bounced off the satellites to the local cable companies or directly to homeowners. Such satellites would likely be targets themselves and are just one of the reasons that some government have been spending plenty of money trying to develop killer satellites.

One of the other dangers of satellite television is that a country could easily be bombarded with television signals from a foreign government's satellites. That happens to a certain extent between the United States and Canada, but it isn't too great a problem because the two countries get along reasonably well by international standards.

What about the United States and Cuba or some of our South American and Central American neighbors, where "Yankee" is a dirty word? Do you think that the people in those countries appreciate being bombarded with American programming, simply because American firms have the money and know-now to do so? Imagine how Americans in south Florida would feel being bombarded by Soviet programming intended for their "friends" in Cuba. If some underdeveloped nation wants to put up a satellite, the fact that all the orbital slots are full will only increase that nation's determination (and rhetoric).

An Incident in Phoenix

Early in 1982 a man barged into a Phoenix television station (KOOL-TV) and received national media attention as he pistol-whipped a station employee and then held the employee at gunpoint while he negotiated for air time. In exchange for the employee's freedom, the station agreed to broadcast a rambling, incoherent message that lasted more than 20 minutes. The news anchorman who read the message had a revolver pointed to his stomach during the entire reading. Upon completion of the newsman's reading, the gunman calmly handed over his gun, shook hands, and surrendered to the Phoenix police department.

While this was hopefully an extremely isolated incident, the fact remains that satellite television systems share some of the same vulnerability. An individual could take over a system for a limited time, causing untold damage or hysteria. If satellite television systems had been widely used at the time of the Patty Hearst kidnapping, would her tormentors have considered demanding free satellite air time? Would they have been successful in their quest?

RELIGION AND SATELLITE COMMUNICATIONS

Among the first groups to begin leasing transponder space were evangelical Christians who saw modern satellite communications as an ideal way to spread their religion. You can find programming for any of the three major faiths found in North America (Catholic, Protestant, and Jewish), and many of the individual denominations to be found within those faiths.

A cynic might ask if God really needs the aid of a satellite to reach out and touch a follower. In reality satellite communications are an excellent medium for religious programming, especially since such programming is frowned upon by the major commercial networks.

With the use of satellites, religious programmers can find a ready outlet for their relatively inexpensive-to-produce programming. They don't have to pay a small fortune to buy commercial network time.

Let's say that you are the driving force behind a small but revolutionary new religion. You have 20,000 followers. Unfortunately, all are within 60 miles of Topeka, Kansas. One way to expand the religion would be the rather slow method of sending out missionaries to other parts of the country. Because the religion is so revolutionary, the concept of missionaries is unthinkable to the church members. They are willing to contribute lots of money to any worthwhile project, but they are unwilling to go more than 75 miles from Topeka.

One way to expand the new religion would be to record a television program, lease some transponder time, and broadcast the program all over North America. If you sign a long-term contract for the same time slot every week or every day, your programming would appear in one of the satellite television programming directories. The converts would begin to roll in.

Let's look at the same scenario. Take away the element of honesty, and assume you are doing this for money. The rules governing content of satellite television programming are virtually nonexistent. Some very nice fund raising scams—all in the name of old-time religion—could be run using modern technology and old-time gullibility.

Appendices

Appendix A

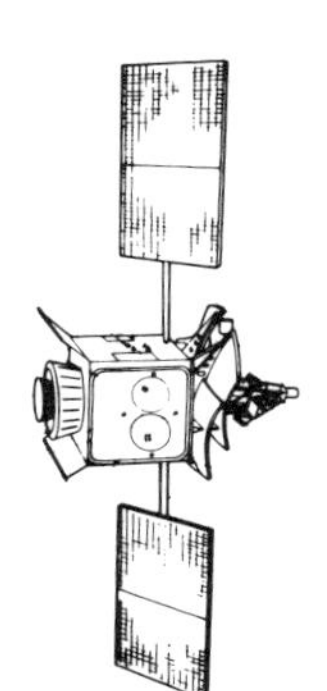

Table of Recent Artificial Satellites Launched

Code name Spacecraft description	International number	Country Organization Site of launching	Date	Perigee Apogee	Period Inclination	Frequencies Transmitter power	Observations
Cosmos-1149	1980-1-A	USSR (PLE)	9 Jan	208 km 414 km	90.4 min 70.29°		Photographic reconnaissance satellite Recovered on 23 January 1980
46th Molnya-1 hermetically sealed cylinder with conical ends, mass: 1000 kg, 6 solar panels	1980-2-A	USSR (PLE)	11 Jan	478 km 40 830 km	737 min 62.8°	800 MHz band 40 W (emission) 1000 MHz band (reception) 3400-4100 MHz (retransmission of television)	Apparatus for transmitting television programmes and multichannel radiocommunications
Cosmos-1150	1980-3-A	USSR (PLE)	14 Jan	989 km 1028 km	105.0 min 83.0°		Navigation satellite
FLTSATCOM-3 3-axis stabilized hexagonal satellite, width: 2.44 m, overall height: 6.70 m, mass at launch: 1875 kg, mass in orbit: 1005 kg	1980-4-A	United States USDOD (ETR)	18 Jan	35 745 km; 35 829 km in geostationary orbit at 23° W	1436.1 min 2.6°	240-400 MHz band (communications) 2252.5; 2262.5 MHz 2.4 W (telemetry)	Government communications satellite. Replaces *FLTSATCOM-2* which is being moved to 75° E. Nine 25 kHz channels and twelve 5 kHz channels for small mobile users. A 25 kHz broadcast channel and a 500 kHz channel
Cosmos-1151	1980-5-A	USSR (PLE)	23 Jan	650 km 678 km	97.8 min 82.5°		Ocean monitoring satellite
Cosmos-1152	1980-6-A	USSR (PLE)	24 Jan	181 km 370 km	89.7 min 67.1°		High-resolution, reconnaissance satellite Recovered on 6 February 1980
Cosmos-1153	1980-7-A	USSR (PLE)	25 Jan	983 km 1031 km	105 min 83°		Navigation satellite
Cosmos-1154	1980-8-A	USSR (PLE)	30 Jan	634 km 671 km	97.3 min 81.3°		Electronic monitoring satellite
Cosmos-1155	1980-9-A	USSR (PLE)	7 Feb.	206 km 422 km	90.4 min 72.9°		Medium-resolution reconnaissance satellite Recovered on 21 February 1980
KH-11	1980-10-A	United States USAF (WTR)	7 Feb.	220 km 498 km	91.7 min 97.0°		Digital imaging reconnaissance satellite

Code name Spacecraft description	International number	Country Organization Site of launching	Date	Perigee Apogee	Period Inclination	Frequencies Transmitter power	Observations
Navstar-5	1980-11-A	United States (WTR)	9 Feb.	20 095 km 20 165 km	715.9 min 63.7°		Global positioning system navigation satellite. Replaces *Navstar-1*
Cosmos-1156 to **Cosmos-1163** mass: 40 kg each	1980-12-A to 1980-12-H	USSR (PLE)	12 Feb.	1450 km 1528 km	115.4 min 74.0°		Government communication satellites
Cosmos-1164	1980-13-A	USSR (PLE)	12 Feb.	220 km 640 km	92.9 min 62.8°		
SMM 3-axis stabilized satellite; width: 1.20 m; length: 4 m; mass: 2315 kg; 2 fixed solar arrays (3 kW); Ni-Cd batteries	1980-14-A	United States NASA (ETR)	14 Feb.	571.5 km 573.5 km	96.12 min 28.5°	2287.5 MHz (tracking and telemetry)	Solar Maximum Mission. Objectives: to measure solar radiation during the period of maximum solar activity. Carries gamma ray spectrometer, hard X-ray burst spectrometer, hard X-ray imaging spectrometer, ultraviolet spectrometer and polarimeter, X-ray polychrometer, coronograph/polarimeter and solar constant monitoring package
Tansei-4 (MS-T4)	1980-15-A	Japan ISAS (KSC)	17 Feb.	517 km 672 km	96.5 min 38.7°	136.725 MHz (tracking) 400.45; 2280.5 MHz (telemetry)	Tests of new technology for future satellites and test of the new *M-3S* launcher
Raduga-5 3-axis stabilized satellite; mass: 5 tonnes; solar cells	1980-16-A	USSR (BAI)	20 Feb.	36 610 km geosynchronous orbit	24 h 38 min 0.4°	5.7-6.2 GHz (reception) 3.4-3.9 GHz (emission)	Carries apparatus for transmitting television programmes and multichannel radiocommunications
Cosmos-1165	1980-17-A	USSR (PLE)	21 Feb.	182 km 379 km	89.8 min 72.9°		High-resolution reconnaissance satellite. Recovered on 5 March 1980

Code name Spacecraft description	International number	Country Organization Site of launching	Date	Perigee Apogee	Period Inclination	Frequencies Transmitter power	Observations
Ayame-2 cylindrical satellite; diameter: 1 m; height: 1.50 m; mass: 260 kg	1980-18-A	Japan NSDA (TSC)	22 Feb.	206.9 km 35 512 km	625.8 min 24.6	31.65 GHz 3.2 W 4.075; 4.080 GHz 4.7 W 3.940 GHz 3.5 W 136.112 MHz 2 or 8 W	Experimental telecommunication satellite. Was intended for geostationary orbit but contact was lost while still in transfer orbit
No name	1980-19-A to 1980-19-C	United States USN (WTR)	3 March	1053 km 1151 km	107.1 min 63.5°		Ocean surveillance satellite system. Three satellites
Cosmos-1166	1980-20-A	USSR (PLE)	4 March	208 km 406 km	90.3 min 72.9		Medium-resolution photographic reconnaissance satellite. Recovered on 18 March 1980
Cosmos-1167	1980-21-A	USSR (PLE)	14 March	438 km 457 km	93.3 min 65°		Ocean surveillance satellite
Cosmos-1168	1980-22-A	USSR (PLE)	17 March	981 km 1026 km	104.9 min 82.9°		Navigation satellite
Cosmos-1169	1980-23-A	USSR (PLE)	27 March	478 km 521 km	94.5 min 65.8°		Satellite intercept programme
Progress-8 modified *Soyuz* spacecraft without the descent section; mass at launch: 7 tonnes	1980-24-A	USSR (BAI)	27 March	192 km 266 km	88.8 min 51.6°		Expendable supply craft. Docked with *Salyut-6* on 29 March. Separated on 25 April and was deorbited over the Pacific Ocean on 26 April 1980
Cosmos-1170	1980-25-A	USSR (BAI)	1 April	181 km 386 km	89.9 min 70.4°		High-resolution photographic reconnaissance satellite. Recovered on 13 April 1980
Cosmos-1171	1980-26-A	USSR (PLE)	3 April	976 km 1017 km	105 min 65.8°		Satellite intercept programme. Target vehicle for *Cosmos-1174*
Soyuz-35 3-part spacecraft: 2 spherical habitable modules (orbital compartment and command module) connected in tandem to a cylindrical service module; diameter: 2.70 m; height: 7.10 m; mass: 6.7 tonnes; 2 solar arrays	1980-27-A	USSR (BAI)	9 April	276 km 315 km	90.3 min 51.6°		Two-man spacecraft. L. Popov, commander; V. Ryumin, flight engineer. Docked with *Salyut-6* on 10 April. Returned to Earth carrying *Soyuz-36* cosmonauts on 3 June 1980, landing some 440 km north-east of Baikonur Cosmodrome

Cosmos-1172	1980-28-A	USSR (PLE)	12 April	637 km 40 160 km	726 min 62.8°		Early warning satellite
Cosmos-1173	1980-29-A	USSR (BAI)	17 April	180 km 379 km	89.9 min 70.3°		High-resolution photographic reconnaissance satellite. Similar to *Cosmos-1170*. Recovered on 28 April 1980
Cosmos-1174	1980-30-A	USSR (BAI)	18 April	387 km 1035 km	98.6 min 65.8°		Satellite intercept programme. Interceptor vehicle for *Cosmos-1171* target vehicle. The test was a failure. *Cosmos-1174* was exploded in space on 20 April 1980
Cosmos-1175	1980-31-A	USSR (PLE)	18 April	317 km 485 km	92.3 min 62.5°		Satellite intercept programme. Interceptor vehicle employing an optical-thermal guidance system. Decayed on 28 May 1980
Navstar-6	1980-32-A	United States (WTR)	26 April	19 622 km 20 231 km	707.6 min 62.9°		Global positioning system navigation satellite
Progress-9 modified *Soyuz* spacecraft without the descent section; mass at launch: 7 tonnes	1980-33-A	USSR (BAI)	27 April	192 km 275 km	88.9 min 51.6°		Expendable supply craft. Docked with *Salyut-6* on 29 April, undocked on 20 May and was made to re-enter the Earth's atmosphere on 22 May 1980
Cosmos-1176	1980-34-A	USSR	29 April	260 km 265 km	89.6 min 65.0°		Ocean surveillance satellite. Carries a nuclear reactor. Similar to *Cosmos-954*
Cosmos-1177	1980-35-A	USSR (PLE)	29 April	181 km 365 km	89.7 min 67.2°		High-resolution photographic reconnaissance satellite. Recovered on 12 June 1980
Cosmos-1178	1980-36-A	USSR (PLE)	7 May	207 km 417 km	90.4 min 72.9°		Reconnaissance satellite. Recovered on 22 May 1980
Cosmos-1179	1980-37-A	USSR (PLE)	14 May	310 km 1570 km	103.5 min 83.0°		Navigation satellite
Cosmos-1180	1980-38-A	USSR (PLE)	15 May	240 km 296 km	89.8 min 62.8°		Satellite for geophysical observations and measurements. Recovered on 25 May 1980

Code name *Spacecraft description*	*International number*	*Country* *Organization* *Site of launching*	*Date*	*Perigee* *Apogee*	*Period* *Inclination*	*Frequencies* *Transmitter power*	*Observations*
Cosmos-1181	1980-39-A	USSR	20 May	992 km 1020 km	105 min 82°		Navigation satellite
Cosmos-1182	1980-40-A	USSR (PLE)	23 May	221 km 278 km	89.2 min 82.3°		Medium-resolution photographic reconnaissance satellite. Recovered on 5 June 1980
Soyuz-36 3-part spacecraft: 2 spherical habitable modules (orbital compartment and command module) connected in tandem to a cylindrical service module; diameter: 2.70 m; height: 7.10 m; mass: 6.7 tonnes; 2 solar arrays	1980-41-A	USSR (BAI)	26 May	198 km 216 km	88.0 min 51.6°		Two-man spacecraft. V. Kubasov, commander; B. Farkas (Hungary), research cosmonaut. Docked with *Salyut-6* on 28 May. *Soyuz-36* was returned to Earth with *Soyuz-37* cosmonauts aboard on 31 July 1980
Cosmos-1183	1980-42-A	USSR	28 May	208 km 414 km	90.4 min 72.9°		Medium-resolution photographic reconnaissance satellite. Recovered on 11 June 1980
NOAA-B	1980-43-A	United States NOAA (WTR)	29 May	273 km 1453 km	102.2 min 92.3°		Owing to malfunction of the launch vehicle, proper orbit was not attained and the spacecraft is considered inoperable
Cosmos-1184	1980-44-A	USSR (PLE)	4 June	621 km 662 km	97.4 min 81.2°		Electronic monitoring satellite
Soyuz-T2 solar batteries	1980-45-A	USSR (BAI)	5 June	267 km 316 km	90.25 min 51.6°		Two-man spacecraft. Y. Malishev and V. Aksenov, cosmonauts. Docked with accessories port of *Salyut-6* on 6 June. Recovered on 9 June 200 km south-east of Dzhezkazgan
Cosmos-1185	1980-46-A	USSR (PLE)	6 June	226 km 308 km	89.5 min 82.3°		Medium-resolution photographic reconnaissance satellite. Recovered on 20 June 1980
Cosmos-1186	1980-47-A	USSR (PLE)	6 June	473 km 519 km	94.5 min 74.0°		Electronic monitoring satellite

Cosmos-1187	**1980-48-A**	**USSR (PLE)**	**12 June**	**210 km 332 km**	**72.9 min 89.6°**		**Medium-resolution photographic reconnaissance satellite.** **Recovered on 26 June 1980**
Gorizont-4 3-axis stabilized spacecraft	1980-49-A	USSR (BAI)	14 June	36 515 km	24 h 33 min 0.8°	3.4-3.9 GHz (emission) 5.7-6.2 GHz (reception)	Communication satellite for transmission of telegraph and telephone messages and for transmission of television programmes
Cosmos-1188	1980-50-A	USSR (PLE)	14 June	628 km 40 165 km	726 min 62.8°		Early warning satellite
30th Meteor-1 3-axis stabilized cylindrical satellite mass: 2200 kg; sun-oriented solar panels	1980-51-A	USSR CAHS (PLE)	18 June	589 km 678 km	97.3 min 98.0°		Meteorological satellite
Big Bird	1980-52-A	United States USAF (WTR)	18 June	165 km 254 km	88.5 min 96.5°		Reconnaissance satellite
No name	1980-52-C	United States	18 June	1325 km 1329 km	112.2 min 96.6°		
47th Molnya-1 hermetically-sealed cylinder with conical ends; mass: 1000 kg; 6 solar panels	1980-53-A	USSR (PLE)	21 June	658 km 40 707 km	738 min 62.5°	800 MHz band 40 W (emission) 1000 MHz band (reception) 3400-4100 MHz (retransmission of television)	Television and multichannel radiocommunications
Cosmos-1189	1980-54-A	USSR (PLE)	26 June	209 km 330 km	89.5 min 72.9°		Medium-resolution photographic reconnaissance satellite. Recovered on 10 July 1980
Progress-10 modified *Soyuz* spacecraft without the descent section; mass at launch: 7 tonnes	1980-55-A	USSR	29 June	191 km 281 km	88.9 min 51.6°		Expendable supply craft. Docked with *Salyut-6* on 1 July and was made to re-enter the Earth's atmosphere on 19 July 1980

Code name Spacecraft description	International number	Country Organization Site of launching	Date	Perigee Apogee	Period Inclination	Frequencies Transmitter power	Observations
Cosmos-1190	1980-56-A	USSR (PLE)	1 July	792 km 829 km	100.8 min 74.0°		Electronic monitoring satellite
Cosmos-1191	1980-57-A	USSR (PLE)	2 July	646 km 40 165 km	726 min 62.8°		Early warning satellite
Cosmos-1192 **to** **Cosmos-1199** mass: 40 kg each	1980-58-A to 1980-58-H	USSR (PLE)	9 July	1451 km 1522 km	115.3 min 74.0°		Government communication satellites
Cosmos-1200	1980-59-A	USSR (PLE)	9 July	209 km 332 km	89.5 min 72.9°		Medium-resolution photographic reconnaissance satellite. Recovered on 23 July 1980
Ekran-5 (Statsionar) 3-axis stabilized satellite; mass: 5 tonnes; solar cells	1980-60-A	USSR (BAI)	14 July	34 474 km geostationary orbit	1420 min 0.36°	5.7-6.2 GHz (reception) 3.4-3.9 GHz (emission)	Television relay satellite
Cosmos-1201	1980-61-A	USSR (PLE)	15 July	220 km 274 km	89.1 min 82.3°		Natural resources satellite. Recovered on 28 July 1980
Rohini-1 mass: 35 kg	1980-62-A	India (SSC)	18 July				First satellite launched by Indian *SLV-3* solid propellant 4-stage rocket system
13th Molnya-3 3-axis stabilized satellite; mass: 1500 kg	1980-63-A	USSR (PLE)	18 July	467 km 40 815 km	736 min 62.8°	5.9-6.2 GHz (reception) 3.6-3.9 GHz (emission)	Television and multichannel radiocommunications
Soyuz-37 3-part spacecraft: 2 spherical habitable modules (orbital compartment and command module) connected in tandem to a cylindrical service module; diameter: 2.70 m; height: 7.10 m; mass: 6.7 tonnes; 2 solar arrays	1980-64-A	USSR (BAI)	23 July	263 km 312 km	90.0 min 51.6°		Two-man spacecraft: cosmonaut V. Gorbatko and cosmonaut-researcher Fam Tuan (Viet Nam). Docked with *Salyut-6* on 24 July. *Soyuz-37* cosmonauts returned to Earth aboard *Soyuz-36* on 31 July 1980. Soyuz-37 spacecraft was returned to Earth with *Soyuz-35* cosmonauts Popov and Ryumin on 11 October 1980

Cosmos-1202	1980-65-A	USSR (PLE)	25 July	209 km 333 km	89.6 min 72.9°		Medium-resolution photographic reconnaissance satellite. Recovered on 7 August 1980
Cosmos-1203	1980-66-A	USSR (PLE)	31 July	227 km 303 km	89.5 min 82.3°		Recovered on 14 August 1980
Cosmos-1204	1980-67-A	USSR (AKY)	31 July	346 km 546 km	93.3 min 50.7°		
Cosmos-1205	1980-68-A	USSR (PLE)	12 Aug.	208 km 332 km	89.6 min 72.8°		Photographic reconnaissance satellite. Recovered on 26 August 1980
Cosmos-1206	1980-69-A	USSR (PLE)	15 Aug.	630 km 659 km	97.4 min 81.2°		Electronic monitoring satellite
Cosmos-1207	1980-70-A	USSR (PLE)	22 Aug.	218 km 282 km	89.2 min 82.3°		Film-return earth resources satellite. Recovered on 4 September 1980
Cosmos-1208	1980-71-A	USSR	26 Aug.	181 km 362 km	89.6 min 67.1°		Long-duration reconnaissance satellite. Recovered on 24 September 1980
Cosmos-1209	1980-72-A	USSR (PLE)	3 Sept.	222 km 306 km	89.4 min 82.3°		Earth resources satellite. Recovered on 17 September 1980
6th Meteor-2	1980-73-A	USSR CAHS (PLE)	9 Sept.	868 km 906 km	102.4 min 81.2°	137.3 MHz 5 W (APT)	Meteorological satellite. Scanning telephotometer and television-type scanning equipment (0.5 to 0.7 μm), infrared scanning radiometer (8 to 12 μm)
GOES-4 cylindrical spin-stabilized satellite: diameter: 1.90 m; height: 2.30 m; mass: 397 kg	1980-74-A	United States (ETR)	9 Sept.	34 264 km 49 830 km	1767 min 0.25° in geosynchronous orbit at 95° W	2209 MHz; 2214 MHz (telemetry)	Geostationary Operational Environmental Satellite. Carries a visible and infrared spin-scan radiometer (VISSR) to provide data on the vertical structures of temperature and moisture in the atmosphere

Code name Spacecraft description	International number	Country Organization Site of launching	Date	Perigee Apogee	Period Inclination	Frequencies Transmitter power	Observations
Soyuz-38 3-part spacecraft: 2 spherical habitable modules (orbital compartment and command module) connected in tandem to a cylindrical service module: diameter: 2.70 m: height: 7.10 m: mass: 6680 kg: 2 solar arrays	1980-75-A	USSR (BAI)	18 Sept.	199 km 273 km	88.9 min 51.6°		Two-man spacecraft: Y. V. Romanenko, flight commander: A. Tomayo Méndez (Cuba). Docked with *Salyut-6* on 19 September 1980 and returned to Earth with the same crew on 26 September 1980
Cosmos-1210	1980-76-A	USSR (PLE)	19 Sept.	195 km 268 km	88.8 min 82.3°		Photographic reconnaissance satellite. Recovered on 30 September 1980
Cosmos-1211	1980-77-A	USSR (PLE)	23 Sept.	215 km 261 km	89.1 min 82.4°		Photographic reconnaissance satellite. Recovered on 4 October 1980
Cosmos-1212	1980-78-A	USSR (PLE)	26 Sept.	216 km 275 km	89.1 min 82.3°		Earth resources satellite. Recovered on 9 October 1980
Progress-11 modified *Soyuz* spacecraft without the descent section: mass at launch: 7 tonnes	1980-79-A	USSR (BAI)	28 Sept.	193 km 270 km	88.8 min 51.6°		Cargo-spacecraft. Docked with the *Salyut-6/Soyuz-37* complex on 30 September 1980. Made to re-enter the Earth's atmosphere on 11 December 1980
Cosmos-1213	1980-80-A	USSR (PLE)	3 Oct.	207 km 343 km	89.6 min 72.8°		Photographic reconnaissance satellite Recovered on 17 October 1980
Raduga-6 (Statsionar-3) 3-axis stabilized satellite: mass: 5 tonnes: solar cells	1980-81-A	USSR (BAI)	6 Oct.	36 000 km geostationary orbit	1444 min (24 h 04 min) 0.4°	5.7-6.2 GHz (reception) 3.4-3.9 GHz (emission)	Carries apparatus for transmitting television programmes and multichannel radiocommunications
Cosmos-1214	1980-82-A	USSR (PLE)	10 Oct.	181 km 368 km	89.7 min 67.2°		Photographic film recovery reconnaissance satellite. Recovered on 23 October 1980
Cosmos-1215	1980-83-A	USSR (PLE)	14 Oct.	499 km 553 km	95.1 min 74.0°		Electronic monitoring satellite

Cosmos-1216	1980-84-A	USSR (PLE)	16 Oct.	209 km 404 km	90.3 min 72.9°		Photographic film recovery reconnaissance satellite. Recovered on 30 October 1980
Cosmos-1217	1980-85-A	USSR (PLE)	24 Oct.	642 km 40 165 km	726 min 62.8°		Early warning satellite
Cosmos-1218	1980-86-A	USSR (PLE)	30 Oct.	178 km 374 km	89.7 min 64.9°		High-resolution photographic reconnaissance satellite. Recovered on 12 December 1980
FLTSATCOM-4 3-axis stabilized hexagonal satellite; width: 2.44 m; overall height: 6.70 m; mass at launch: 1876 kg; mass in orbit: 1005 kg	1980-87-A	United States USN (ETR)	31 Oct.	35 033 km 36 237 km in geostationary orbit at 172° E	1428.4 min 2.5°	240-400 MHz band (communications) 2252.5; 2262.2 MHz 2.4 W (telemetry)	Government communication satellite providing 23 UHF communication channels and one SHF up-link channel. Fourth in a series of five satellites
Cosmos-1219	1980-88-A	USSR (PLE)	31 Oct.	205 km 353 km	89.7 min 72.9°		Medium-resolution photographic reconnaissance satellite. Recovered on 13 November 1980
Cosmos-1220	1980-89-A	USSR (BAI)	4 Nov.	432 km 454 km	93.3 min 65.0°		Ocean surveillance satellite
Cosmos-1221	1980-90-A	USSR (PLE)	12 Nov.	207 km 424 km	90.5 min 72.9°		Medium-resolution photographic reconnaissance satellite
SBS-1 mass: 550 kg	1980-91-A	United States SBS (ETR)	15 Nov.	in geostationary orbit at 106° W		14-12 GHz band	United States domestic communication satellite. First of three all-digital business communications satellites. Transmits point-to-point voice, data, facsimile and telex messages. Ten transponders
48th Molnya-1 hermetically sealed cylinder with conical ends; mass: 1000 kg; 6 solar panels	1980-92-A	USSR (PLE)	16 Nov.	640 km 40 651 km	736 min 62.3°	800 MHz band 40 W (emission) 1000 MHz band (reception) 3400-4100 MHz (retransmission of television)	Television and multichannel radiocommunications

Cosmos-1222	1980-93-A	USSR (PLE)	21 Nov.	624 km 659 km	97.4 min 81.2°		Electronic monitoring satellite
Soyuz-T3	1980-94-A	USSR (BAI)	27 Nov.	253 km 271.5 km	89.6 min 51.6°		For the first time in nine years three cosmonauts were launched aboard a *Soyuz*: L. Kizim, flight commander; O. Makarov, flight engineer; G. Strekalov, research engineer. *Soyuz-T3* docked with *Salyut-6* on 28 November and the crew boarded Salyut-6 on 29 November Soyuz-T3 was returned to Earth with its crew on 10 December, landing in Kazakhstan
Cosmos-1223	1980-95-A	USSR (PLE)	27 Nov.	614 km 40 165 km	726 min 62.8°		Early warning satellite
Cosmos-1224	1980-96-A	USSR (PLE)	1 Dec.	209 km 403 km	90.3 min 72.9°		Medium-resolution photographic reconnaissance satellite Recovered on 15 December 1980
Cosmos-1225	1980-97-A	USSR (PLE)	5 Dec.	967 km 1041 km	105.0 min 82.9°		Navigation satellite
Intelsat-V F2 3-axis stabilized satellite; height: 6.60 m; mass at launch: 1950 kg; 2 solar arrays (1.2 kW)	1980-98-A	International INTELSAT (ETR)	6 Dec.	in geostationary orbit at 335.5° E		2202.5 MHz 3.5 W 5764 MHz 1 W (telemetry) 4-6 GHz (communications)	INTELSAT commercial telecommunication satellite; 12 000 telephone channels and two colour television channels
Cosmos-1226	1980-99-A	USSR (PLE)	10 Dec.	982 km 1025 km	105.0 min 83.0°		Navigation satellite
No name	1980-100-A	United States USAF (WTR)	13 Dec.	250 km 39 127 km	63.8°		Satellite data systems spacecraft. Provides UHF communications and relays data and communications between satellite control facility earth stations
Cosmos-1227	1980-101-A	USSR (PLE)	16 Dec.	209 km 325 km	89.5 min 72.9°		Medium-resolution photographic reconnaissance satellite
Cosmos-1228 **to** **Cosmos-1235** mass: 40 kg each	1980-102-A to 1980-102-H	USSR (PLE)	24 Dec.	1415 km 1491 km	114.6 min 74°		Government communication satellites

Code name *Spacecraft description*	*International number*	*Country* *Organization* *Site of launching*	*Date*	*Perigee* *Apogee*	*Period* *Inclination*	*Frequencies* *Transmitter power*	*Observations*
Prognoz-8 pressurized central body; 4 solar panels	1980-103-A	USSR (BAI)	25 Dec.	550 km 199 000 km	95 h 23 min 65°		Automatic satellite to study influence of solar activity on Earth's magnetosphere. Experiments or equipment have been supplied by Czechoslovakia, Poland and Sweden
Ekran-6 **(Statsionar-T)**	1980-104-A	USSR (BAI)	26 Dec.	35 554 km 35 554 km geostationary orbit	1424 min 0.4°	5.7-6.2 GHz (reception) 3.4-3.9 GHz (emission)	Television relay satellite
Cosmos-1236	1980-105-A	USSR (PLE)	26 Dec.	180 km 388 km	89.8 min 67.1°		High-resolution photographic reconnaissance satellite

Appendix B

Appendix B

Table of Frequency Allocations

§2.106 Table of Frequency Allocations

Worldwide		Region 2		United States	
Band (kHz)	Service	Band (kHz)	Service	Band (kHz)	Allocation
1	2	3	4	5	6
Below 10 (157)	(Not allocated).			Below 10	(Not allocated).
10–14	RADIONAVIGATION. Radiolocation.				
14–19.95 (159)	FIXED. MARITIME MOBILE. (158)				
19.95–20.05 (159)	STANDARD FREQUENCY. (160)				
20.05–70 (159) (161)	FIXED. MARITIME MOBILE. (158)				
70–90 (161)		70–90 (164)	FIXED. MARITIME MOBILE. (158) MARITIME RADIONAVIGATION. (162) Radiolocation.		
90–110		90–110 (166) (167)	RADIONAVIGATION. FIXED. Maritime mobile. (158)		
110–130		110–130 (164) (167) (168)	FIXED. MARITIME MOBILE. MARITIME RADIO-NAVIGATION. (162) Radiolocation.		
130–160		130–160 (167)	FIXED. MARITIME MOBILE.		
160–200		160–200	FIXED. (179)		

(courtesy Federal Communications Commission).

Band (kHz)	Service	Class of station	Frequency (kHz)	Nature of SERVICES of stations
7	8	9	10	11
Below 10	(Not allocated).			
10–14	RADIONAVIGATION.	Radionavigation land. Radionavigation mobile.		RADIONAVIGATION.
14–19.95	FIXED.	Fixed.		INTERNATIONAL FIXED PUBLIC.
19.95–20.05	STANDARD FREQUENCY.	Standard frequency.	20	Standard frequency.
20.05–59	FIXED.	Fixed.		INTERNATIONAL FIXED PUBLIC.
59–61	STANDARD FREQUENCY.	Standard frequency.	60	Standard frequency.
61–70	FIXED.	Fixed.		INTERNATIONAL FIXED PUBLIC.
70–90	FIXED. Radiolocation.	Fixed. Radiolocation land. Radiolocation mobile.		INTERNATIONAL FIXED PUBLIC. RADIOLOCATION.
90–110 (166) (US18) (US104)	RADIONAVIGATION.	Radionavigation land. Radionavigation mobile.		RADIONAVIGATION.
110–130	FIXED. MARITIME MOBILE. Radiolocation.	Coast. Fixed. Radiolocation land. Radiolocation mobile. Ship.		FIXED (in Alaska). INTERNATIONAL FIXED PUBLIC. MARITIME MOBILE. RADIOLOCATION.
130–160	FIXED. MARITIME MOBILE.	Coast. Fixed. Ship.		FIXED (in Alaska). INTERNATIONAL FIXED PUBLIC. MARITIME MOBILE.
160–190	FIXED.	Fixed.		FIXED (in Alaska). INTERNATIONAL FIXED PUBLIC.

§ 2.106 Table of Frequency Allocations—Continued

Worldwide		Region 2		United States	
Band (kHz)	Service	Band (kHz)	Service	Band (kHz)	Allocation
1	2	3	4	5	6
200–285		200–285	AERONAUTICAL RADIO-NAVIGATION. Aeronautical mobile.		
285–315	MARITIME RADIO-NAVIGATION (radiobeacons). Aeronautical radio-navigation.				
315–325		315–325	MARITIME RADIO-NAVIGATION (radiobeacons). Aeronautical radio-navigation		
325–405	AERONAUTICAL RADIO-NAVIGATION. Aeronautical mobile. (181)				
405–415 (182)		405–415	MARITIME RADIO-NAVIGATION (radio direction-finding). *Aeronautical radio-navigation.* Aeronautical mobile.		
415–490	MARITIME MOBILE. (185) (186)				

Federal Communications Commission

Band (kHz)	Service	Class of station	Frequency (kHz)	Nature of services of stations
7	8	9	10	11
190–200	AERONAUTICAL RADIONAVIGATION. (US 226)	Radionavigation land.		AERONAUTICAL RADIONAVIGATION.
200–275	AERONAUTICAL RADIO-NAVIGATION. (US18) Aeronautical mobile.	Aeronautical. Aircraft Radionavigation land		AERONAUTICAL MOBILE. AERONAUTICAL RADIO-NAVIGATION
275–285	AERONAUTICAL RADIONAVIGATION. Maritime radionavigation (Radiobeacons). (US18) Aeronautical mobile.	•	•	AERONAUTICAL MOBILE. AERONAUTICAL RADIONAVIGATION. MARITIME RADIO NAVIGATION.
285–325	MARITIME RADIO-NAVIGATION. (US18)	Radionavigation land.		MARITIME RADIO-NAVIGATION.
325–335	AERONAUTICAL RADIONAVIGA- Maritime radionavigation (Radiobeacons). (US18) Aeronautical mobile.	•	•	AERONAUTICAL MOBILE. AERONAUTICAL RADIONAVIGATION. MARITIME RADIO-NAVIGATION.
335–405	AERONAUTICAL RADIO-NAVIGATION. (US18) Aeronautical mobile.	Aeronautical. Aircraft. Radionavigation land.		AERONAUTICAL MOBILE. AERONAUTICAL RADIO-NAVIGATION.
405–415 (US18)	MARITIME RADIO-NAVIGATION (radio direction-finding). Aeronautical radio-navigation. Aeronautical mobile. (US19)	Aeronautical. Aircraft. Radionavigation land. Radionavigation mobile.	410	Radio direction-finding.
415–490	MARITIME MOBILE.	Coast. Ship.		MARITIME MOBILE (telegraphy).

490–510 (187)	MOBILE (distress and calling).				
510–525 (185)		510–525	MOBILE. *Aeronautical radionavigation.* (188)		
525–535		525–535	MOBILE. *Broadcasting.* (191) *Aeronautical radionavigation.* (188)		
535–1605	BROADCASTING.				
1605–1800		1605–1800	FIXED. MOBILE. AERONAUTICAL RADIONAVIGATION. Radiolocation.		
1800–2000		1800–2000 (198)	AMATEUR. FIXED. MOBILE except aeronautical mobile. RADIONAVIGATION.		

490-510	MOBILE	Coast. Mobile.	500	Distress and calling.
510-525 (US14) (US225)	AERONAUTICAL RADIONAVIGATION Maritime radionavigation (Radiobeacons). (US18)	Radionavigation land.		AERONAUTICAL RADIONAVIGATION. MARITIME RADIONAVIGATION.
525-535 (US221)			530	Travelers information.
510-535 (US14) (US221)			530	Travelers information.
535-1605 (US15) (NG16)	BROADCASTING.			Standard broadcasting.
1605-1715 (US97) (US221)	AERONAUTICAL RADIONAVIGATION. FIXED. LAND MOBILE. MARITIME MOBILE. RADIOLOCATION.	Base. Fixed. Land mobile. Radionavigation land. Radiolocation. Coast. Ship.	 1610 1638 1708	AERONAUTICAL FIXED. AERONAUTICAL RADIONAVIGATION. FIXED (in Alaska). INDUSTRIAL. INTERNATIONAL FIXED PUBLIC. MARITIME MOBILE. PUBLIC SAFETY. RADIOLOCATION. Remote pickup broadcast base. Remote pickup broadcast mobile. Travelers information. Radionavigation land. Do.
1715-1750	FIXED. LAND MOBILE. MARITIME MOBILE. RADIOLOCATION.	Base. Mobile. Fixed. Land mobile. Ship.		AERONAUTICAL FIXED. FIXED (in Alaska). INDUSTRIAL. INTERNATIONAL FIXED PUBLIC. MARITIME MOBILE. PUBLIC SAFETY. RADIOLOCATION.
1750-1800 (NG14)	FIXED. MOBILE. RADIOLOCATION.	Fixed. Land. Mobile.		DISASTER. RADIOLOCATION.
1800-2000 (198) (US18)	RADIONAVIGATION. Amateur. (NG15)	Amateur. Loran.		AMATEUR. Loran

§ 2.106 Table of Frequency Allocations—Continued

Worldwide		Region 2		United States	
Band (kHz)	Service	Band (kHz)	Service	Band (kHz)	Allocation
1	2	3	4	5	6
2000–2065		2000–2065	FIXED. MOBILE		
2065–2107		2065–2107	MARITIME MOBILE. (200)		
2107–2170		2107–2170	FIXED. MOBILE.		
2170–2194	MOBILE (distress and calling). (201) (201A)				

Band (kHz) 7	Service 8	Class of station 9	Frequency (kHz) 10	Nature of stations / OF SERVICES 11
2000–2035	MARITIME MOBILE. (NG19)	Coast. Ship.		MARITIME MOBILE.
2035–2065	MARITIME MOBILE.	Coast.		Coast (telegraphy).
2065–2068.5 (200)	MARITIME MOBILE.	Coast. Ship.		MARITIME MOBILE (telephony).
2068.5–2078.5	MARITIME MOBILE.	Ship.		Ship (wideband telegraphy, facsimile, and special transmission systems).
2078.5–2089.5 (200)	MARITIME MOBILE.	Coast. Ship.		MARITIME MOBILE (telephony).
2089.5–2092.5	MARITIME MOBILE.	Ship.		Ship (calling, telegraphy).
2092.5–2107 (200)	MARITIME MOBILE.	Coast. Ship.		MARITIME MOBILE.
2107–2170	FIXED. LAND MOBILE. MARITIME MOBILE. (NG19)	Base. Coast. Fixed. Land mobile. Ship.		AERONAUTICAL FIXED. FIXED (in Alaska). INDUSTRIAL. INTERNATIONAL FIXED PUBLIC. MARITIME MOBILE. PUBLIC SAFETY.
2170–2173.5	MARITIME MOBILE.	Ship.		MARITIME MOBILE.
2173.5–2190.5 (US113)	MOBILE.	Aircraft. Coast. Ship. Survival craft.	 2182	AERONAUTICAL MOBILE (telephony). MARITIME MOBILE (telephony). (NG22) Distress and calling frequency.
2190.5–2194	MARITIME MOBILE.	Ship.		MARITIME MOBILE.

2194–2300		2194–2300	FIXED. MOBILE.		
2300–2495		2300–2495	FIXED. MOBILE. BROADCASTING. (202)		
2495–2505		2495–2505 (203) (203A)	STANDARD FREQUENCY.		
2505–2625		2505–2625	FIXED. MOBILE.		
2625–2850		2625–2850	FIXED. MOBILE.		
2850–3025	AERONAUTICAL MOBILE. (R)(201A)				
3025–3155	AERONAUTICAL MOBILE. (OR)				
3155–3200	FIXED. MOBILE except aeronautical mobile. (R)				
3200–3230	FIXED. MOBILE except aeronautical mobile. (R) BROADCASTING. (202)				

2194–2495	FIXED. LAND MOBILE. MARITIME MOBILE. (NG19)	Base. Coast. Fixed. Land mobile. Ship.		AERONAUTICAL FIXED. FIXED (in Alaska). INDUSTRIAL. INTERNATIONAL FIXED PUBLIC. MARITIME MOBILE. PUBLIC SAFETY.
2495–2505	STANDARD FREQUENCY.	Standard frequency.	2500	Standard frequency.
2505–2850 (NG20)	FIXED. LAND MOBILE. MARITIME MOBILE.	Base. Coast. Fixed. Land mobile. Ship.	 2638 2738 2804 2808 2812	AERONAUTICAL FIXED. FIXED (in Alaska). INDUSTRIAL. INTERNATIONAL FIXED PUBLIC. MARITIME MOBILE. Intership (telephony). (NG44) Do. PUBLIC SAFETY. Zone and interzone police. Do. Do.
2850–3025 (NG61)	AERONAUTICAL MOBILE. (R)	Aeronautical. Aircraft.		AERONAUTICAL MOBILE.
3025–3155	AERONAUTICAL MOBILE. (OR)	Aeronautical. Aircraft.		AERONAUTICAL MOBILE.
3155–3200	FIXED. LAND MOBILE. MARITIME MOBILE.	Base. Coast. Fixed. Land mobile. Ship.		AERONAUTICAL FIXED. FIXED (in Alaska and Puerto Rico). INDUSTRIAL. INTERNATIONAL FIXED PUBLIC. MARITIME MOBILE. PUBLIC SAFETY.
3200–3230 (NG20)	FIXED. LAND MOBILE. MARITIME MOBILE.	Base. Coast. Fixed. Land mobile. Ship.		AERONAUTICAL FIXED. FIXED (in Alaska). INDUSTRIAL. INTERNATIONAL FIXED PUBLIC. MARITIME MOBILE. PUBLIC SAFETY.

Worldwide		Region 2		United States	
Band (kHz)	Service	Band (kHz)	Service	Band (kHz)	Allocation
1	2	3	4	5	6
3230–3400	FIXED. MOBILE except aeronautical mobile. BROADCASTING. (202)				
3400–3500	AERONAUTICAL MOBILE. (R)				
3500–4000		3500–4000	AMATEUR. FIXED. MOBILE except aeronautical mobile. (R)		
4000–4063	FIXED.				
4063–4438	MARITIME MOBILE. (208) (209)				

Band (kHz) 7	Service 8	Class of station 9	Frequency (kHz) 10	Nature of SERVICES of stations 11
3230–3240 (US105)	FIXED. LAND MOBILE. MARITIME MOBILE.	Base. Coast. Fixed. Land mobile. Ship.		AERONAUTICAL FIXED. FIXED (in Alaska) INDUSTRIAL. INTERNATIONAL FIXED PUBLIC. MARITIME MOBILE. PUBLIC SAFETY.
3240–3400 (US105)	FIXED. LAND MOBILE. MARITIME MOBILE. MOBILE.	Base. Coast. Fixed. Land. Land mobile. Mobile. Ship.		AVIATION (flight test and aeronautical fixed only). FIXED (in Alaska). INDUSTRIAL. INTERNATIONAL FIXED PUBLIC. MARITIME MOBILE. PUBLIC SAFETY.
3400–3500 (NG61)	AERONAUTICAL MOBILE. (R)	Aeronautical. Aircraft.		AERONAUTICAL MOBILE.
3500–4000	AMATEUR.	Amateur.		AMATEUR.
4000–4063	FIXED.	Fixed.		AERONAUTICAL FIXED. FIXED (in Alaska). INTERNATIONAL FIXED PUBLIC.
4063–4143.6	MARITIME MOBILE.	Ship.		Ship (telephony, duplex).
4143.6–4146.6 (US82)	MARITIME MOBILE.	Ship. Coast.	4145	Ship. Coast. (telephony, simplex).
4146.6–4162.5	MARITIME MOBILE.	Ship.		Ship (wide-band telegraphy, facsimile and special transmission systems).
4162.5–4166	MARITIME MOBILE.	Ship. Buoy. Interrogating Coast.		Ship. Buoy. Interrogating Coast. (Oceanographic data transmission).
4166–4170	MARITIME MOBILE.	Ship.	4168	Ship (wide-band telegraphy, facsimile and special transmission systems).

4438-4650		4438-4650	FIXED. MOBILE except aeronautical mobile. (R)		
4650-4700	AERONAUTICAL MOBILE. (R)				
4700-4750	AERONAUTICAL MOBILE. (OR)				
4750-4850		4750-4850	FIXED. BROADCASTING. (202)		
4850-4995	FIXED. LAND MOBILE. BROADCASTING. (202)				

4170–4177.25	MARITIME MOBILE.	Ship.		Ship (paired, narrow-band direct-printing telegraph and data transmission systems, at speeds not exceeding 100 bauds).
4177.25–4179.75	MARITIME MOBILE.	Ship.		Ship (non-pared, narrow-band direct-printing telegraph and data transmission systems, at speeds not exceeding 100 bauds).
4179.75–4187.2	MARITIME MOBILE.	Ship.		Ship (calling, A1 Morse telegraphy).
4187.2–4188	MARITIME MOBILE.	Ship.	4187.6	Ship (digital selective calling).
4188–4219.4	MARITIME MOBILE.	Ship.		Ship (working, A1 Morse telegraphy).
4219.4–4349.4	MARITIME MOBILE.	Coast.		Coast (wide-band and A1 Morse telegraphy, facsimile, special and data transmission systems and direct-printing telegraphy systems).
4349.4–4356.75	MARITIME MOBILE.	Coast.		Coast (paired, narrow-band direct-printing telegraph and data transmissions systems, at speeds not exceeding 100 bauds).
4356.75–4357.4	MARITIME MOBILE.	Coast.	4357	Coast (digital selective calling).
4357.4–4438	MARITIME MOBILE.	Coast.		Coast (telephony, duplex).
4438–4650	**FIXED. MOBILE.**	**Base. Fixed. Mobile.**		**AERONAUTICAL FIXED. FIXED (in Alaska). INDUSTRIAL. INTERNATIONAL FIXED PUBLIC.**
4650–4700 (NG60) (NG61)	**AERONAUTICAL MOBILE. (R)**	**Aeronautical. Aircraft.**		**AERONAUTICAL MOBILE.**
4700–4750	**AERONAUTICAL MOBILE. (OR)**	**Aeronautical. Aircraft.**		**AERONAUTICAL MOBILE.**
4750–4995	**FIXED.**	**Fixed.**		**AERONAUTICAL FIXED. FIXED (in Alaska). INTERNATIONAL FIXED PUBLIC.**

Worldwide		Region 2		United States	
Band (kHz)	Service	Band (kHz)	Service	Band (kHz)	Allocation
1	2	3	4	5	6
4995–5005	STANDARD FREQUENCY. (203A) (210)				
5005–5060	FIXED. BROADCASTING. (202)				
5060–5250	FIXED.				
5250–5450		5250–5450	FIXED. LAND MOBILE.		
5450–5480		5450–5480	AERONAUTICAL MOBILE. (R)		
5480–5680	AERONAUTICAL MOBILE. (R) (201A)				
5680–5730	AERONAUTICAL MOBILE. (OR) (201A)				
5730–5950	FIXED				
5950–6200	BROADCASTING.				
6200–6525 (211)	MARITIME MOBILE.				

Band (kHz)	Service	Class of station	Frequency (kHz)	Nature of SERVICES of stations
7	8	9	10	11
4995–5005	STANDARD FREQUENCY.	Standard frequency.	5000	Standard frequency.
5005–5450	FIXED.	Fixed.		AERONAUTICAL FIXED. FIXED (in Alaska). INTERNATIONAL FIXED PUBLIC. Zone and Interzone police.
5450–5680 (NG61)	AERONAUTICAL MOBILE. (R)	Aeronautical. Aircraft.		AERONAUTICAL MOBILE.
5680–5730	AERONAUTICAL MOBILE. (OR)	Aeronautical. Aircraft.		AERONAUTICAL MOBILE.
5730–5950	FIXED.	Fixed.		AERONAUTICAL FIXED. FIXED (in Alaska). INTERNATIONAL FIXED PUBLIC.
5950–6200 (NG25)	BRAODCASTING.	International broadcasting.		International broadcasting.
6200–6218.6	MARITIME MOBILE.	Ship.		Ship (telephony, duplex).
6218.6–6224.6 (US82)	MARITIME MOBILE.	Ship. Coast.		Ship. Coast. (telephony, simplex).
6224.6–6244.5	MARITIME MOBILE.	Ship.		Ship (side-band telegraphy, facsimile and special transmission systems).
6244.5–6248	MARITIME MOBILE.	Ship. Buoy. Interrogating Coast.		Ship. Buoy. Interrogating Coast. (oceanographic data transmission).
6248–6256	MARITIME MOBILE.	Ship.		Ship (wide-band telegraphy, facsimile and special transmission systems).

6525–6685	AERONAUTICAL MOBILE. (R)				
6685–6765	AERONAUTICAL MOBILE. (OR)				
6765–7000	FIXED.				
7000–7100	AMATEUR. AMATEUR-SATELLITE.				
7100–7300		7100–7300	AMATEUR.		

6256–6267.75	MARITIME MOBILE.	Ship.		Ship (paired, narrow-band direct-printing telegraph and data transmission systems, at speeds not exceeding 100 bauds).
6267.75–6269.75	MARITIME MOBILE.	Ship.		Ship (non-paired, narrow-band direct printing telegraph and data transmission systems, at speeds not exceeding 100 bauds).
6269.75–6280.8	MARITIME MOBILE.	Ship.		Ship (calling, A1 Morse telegraphy).
6280.8–6282	MARITIME MOBILE.	Ship.	6281.4	Ship (digital selective calling).
6282–6325.4	MARITIME MOBILE.	Ship.		Ship (working, A1 Morse telegraphy).
6325.4–6493.9	MARITIME MOBILE.	Coast.		Coast (wide-band and, A1 Morse telegraphy facsimile, special and data transmission systems and direct-printing telegraphy system). (NG27)
6493.9–6505.75	MARITIME MOBILE.	Coast.		Coast (paired, narrow-band direct-printing telegraph and data transmission system, at speeds not exceeding 100 bauds).
6505.75–6506.4	MARITIME MOBILE.	Coast.	6506	Coast (digital selective calling).
6506.4–6525	MARITIME MOBILE.	Coast.		Coast (telephony, duplex)
6525–6685 (NG61)	AERONAUTICAL MOBILE. (R)	Aeronautical. Aircraft.		AERONAUTICAL MOBILE.
6685–6765	AERONAUTICAL MOBILE. (OR)	Aeronautical. Aircraft.		AERONAUTICAL MOBILE.
6765–7000	FIXED.	Fixed.		AERONAUTICAL FIXED. FIXED (in Alaska). INTERNATIONAL FIXED PUBLIC.
7000–7100	AMATEUR. AMATEUR-SATELLITE.	Amateur. Earth. Space.		AMATEUR. AMATEUR-SATELLITE. (NG62)
7100–7300	AMATEUR.	Amateur.		AMATEUR. (NG62)

Worldwide		Region 2		United States	
Band (kHz)	Service	Band (kHz)	Service	Band (kHz)	Allocation
1	2	3	4	5	6
7300–8195	FIXED.				
8195–8815	MARITIME MOBILE. (201A) (213)				

Band (kHz) 7	Service 8	Class of station 9	Frequency (kHz) 10	Nature of services of stations 11
7300-8195	FIXED.	Fixed.		AERONAUTICAL FIXED. FIXED (in Alaska). INTERNATIONAL FIXED PUBLIC. Zone and interzone police.
8195-8291.1	MARITIME MOBILE.	Ship.		Ship (telephony, duplex).
8291.1-8297.3 (US 82)	MARITIME MOBILE.	Ship. Coast.		Ship. Coast (telephony, simplex).
8297.3-8300	MARITIME MOBILE.	Ship.		Ship (non-paired, narrow-band direct-printing telegraph and data transmission systems, at speeds not exceeding 100 bauds).
8300-8328	MARITIME MOBILE.	Ship.		Ship (wide-band telegraphy, facsimile and special transmission systems).
8328-8331.5	MARITIME MOBILE.	Ship. Buoy. Interrogating Coast.		Ship. Buoy. Interrogating Coast (oceanographic data transmission).
8331.5-8343.5	MARITIME MOBILE.	Ship.		Ship (wide-band telegraphy, facsimile and special transmission systems).
8343.5-8357.25	MARITIME MOBILE.	Ship.		Ship (paired, narrow-band direct-printing telegraph and data transmission systems, at speeds not exceeding 100 bauds).
8357.25-8357.75	MARITIME MOBILE.	Ship.	8357.5	Ship (non-paired, narrow-band direct-printing telegraph and data transmission systems, at speeds not exceeding 100 bauds).

8815–8965	AERONAUTICAL MOBILE. (R)				
8965–9040	AERONAUTICAL MOBILE. (OR)				
9040–9500	FIXED.				
9500–9775	BROADCASTING.				
9775–9995	FIXED.				
9995–10005	STANDARD FREQUENCY (201A) (203A) (214)				
10005–10100	AERONAUTICAL MOBILE. (R (201A)				
10100–11175	FIXED.				

8357.75–8359.75	MARITIME MOBILE.	Ship.		Ship (Working, A1 Morse telegraphy).
8359.75–8374.4	MARITIME MOBILE.	Ship.		Ship (calling, A1 Morse telegraphy).
8374.4–8376	MARITIME MOBILE.	Ship.	8375.2	Ship (digital selective calling).
8376–8435.4	MARITIME MOBILE.	Ship.		Ship (working, A1 Morse telegraphy).
8435.4–8704.4	MARITIME MOBILE.	Coast.		Coast (wide-band and A1 Morse telegraphy, facsimile, special and data transmission systems and direct-printing telegraphy systems).
8704.4–8718.25	MARITIME MOBILE.	Coast.		Coast (paired, narrow-band direct-printing telegraph and data transmission systems, at speeds not exceeding 100 bauds).
8718.25–8718.9	MARITIME MOBILE.	Coast.	8718.5	Coast (digital selective calling).
8718.9–8815	MARITIME MOBILE.	Coast.		Coast (telephony, duplex).
8815–8965	AERONAUTICAL MOBILE. (R)	Aeronautical. Aircraft.		AERONAUTICAL MOBILE.
8965–9040	AERONAUTICAL MOBILE. (OR)	Aeronautical. Aircraft.		AERONAUTICAL MOBILE.
9040–9500	FIXED.	Fixed.		AERONAUTICAL FIXED. FIXED (in Alaska). INTERNATIONAL FIXED PUBLIC.
9500–9775	BROADCASTING.	International broadcasting.		International broadcasting.
9775–9995	FIXED.	Fixed.		AERONAUTICAL FIXED. INTERNATIONAL FIXED PUBLIC.
9995–10005 (US204)	STANDARD FREQUENCY.	Standard frequency.	10000	Standard frequency.
10005–10100 (NG61)	AERONAUTICAL MOBILE. (R)	Aeronautical. Aircraft.		AERONAUTICAL MOBILE.
10100–11175	FIXED.	Fixed.		AERONAUTICAL FIXED. INTERNATIONAL FIXED PUBLIC.

Worldwide		Region 2		United States	
Band (kHz)	Service	Band (kHz)	Service	Band (kHz)	Allocation
1	2	3	4	5	6
11175-11275	AERONAUTICAL MOBILE. (OR)				
11275-11400	AERONAUTICAL MOBILE. (R)				
11400-11700	FIXED. (216)				
11700-11975	BROADCASTING.				
11975-12330	FIXED.				
12330-13200	MARITIME MOBILE. (213)				

Band (kHz)	Service	Class of station	Frequency (kHz)	Nature of services of stations
7	8	9	10	11
11175–11275	AERONAUTICAL MOBILE. (OR)	Aeronautical. Aircraft.		AERONAUTICAL MOBILE.
11275–11400 (NG61)	AERONAUTICAL MOBILE. (R)	Aeronautical. Aircraft.		AERONAUTICAL MOBILE.
11400-11700	FIXED.	Fixed.		AERONAUTICAL FIXED. INTERNATIONAL FIXED PUBLIC.
11700–11975	BROADCASTING.	International broadcasting.		International broadcasting.
11975–12330	FIXED.	Fixed.		AERONAUTICAL. FIXED. INTERNATIONAL FIXED PUBLIC.
12330–12429.2	MARITIME MOBILE.	Ship.		Ship (telephony, duplex).
12429.2–12439.5 (US82)	MARITIME MOBILE.	Ship. Coast.		Ship. Coast (telephony, simplex).
12439.5–12479.5	MARITIME MOBILE.	Ship.		Ship (wide-band telegraphy, facsimile and special transmission systems).
12479.5–12483	MARITIME MOBILE.	Ship. Buoy. Interrogating Coast.		Ship. Buoy. Interrogating Coast (oceanographic data transmission).
12483–12491	MARITIME MOBILE.	Ship.		Ship (wide-band telegraphy, facsimile and special transmission system).
12491–12519.75	MARITIME MOBILE.	Ship.		Ship (paired, narrow-band direct-Printing telegraph and data transmission systems, at speeds not exceeding 100 bauds).
12519.75–12526.75	MARITIME MOBILE.	Ship.		Ship (nonpaired, narrow-band direct-printing telepgraph and data transmission systems, at speeds not exceeding 100 bauds).

13200-13260	AERONAUTICAL MOBILE. (OR)				
13260-13360	AERONAUTICAL MOBILE. (R)				
13360-14000 (217)	FIXED.				
14000-14250	AMATEUR. AMATEUR-SATELLITE.				
14250-14350	AMATEUR. (218)				
14350-14990	FIXED.				
14990-15010	STANDARD FREQUENCY. (201A) (203A) (219)				
15010-15100	AERONAUTICAL MOBILE. (OR)				
15100-15450	BROADCASTING.				

12539.6–12561.6	MARITIME MOBILE.	Ship.		Ship (calling, A1 Morse telegraphy).
12561.6–12564	MARITIME MOBILE.	Ship.	12562.3 12562.8	Ship (digital selective calling).
12564–12652.3	MARITIME MOBILE.	Ship.		Ship (working, A1 Morse telegraphy)
12652.3–13070.8	MARITIME MOBILE.	Coast.		Coast (wide-band and A1 Morse telegraphy, facsimile, special and data transmission systems and direct-printing telegraphy systems).
13070.8–13099.75	MARITIME MOBILE.	Coast.		Coast (paired, narrow-band direct-printing telegraph and data transmission systems, at speeds not exceeding 100 bauds)
13099.75–13100.8	MARITIME MOBILE.	Coast.	13100 13100.5	Coast (digital selective calling).
13100.8–13200	MARITIME MOBILE.	Coast.		Coast (telephony, duplex).
13200–13260	AERONAUTICAL MOBILE. (OR)	Aeronautical. Aircraft.		AERONAUTICAL MOBILE.
13260–13360 (NG61)	AERONAUTICAL MOBILE. (R)	Aeronautical. Aircraft.		AERONAUTICAL MOBILE.
13360–14000 (217)	FIXED.	Fixed.	 13560	AERONAUTICAL FIXED. INTERNATIONAL FIXED PUBLIC. Industrial, scientific and medical equipment.
14000–14250	AMATEUR. AMATEUR-SATELLITE.	Amateur. Earth. Space.		AMATEUR. AMATEUR-SATELLITE.
14250–14350	AMATEUR.	Amateur.		AMATEUR.
14350–14990	FIXED.	Fixed.		AERONAUTICAL FIXED. INTERNATIONAL FIXED PUBLIC.
14990–15010 (US204)	STANDARD FREQUENCY.	Standard frequency.	15000	Standard frequency
15010–15100	AERONAUTICAL MOBILE. (OR)	Aeronautical. Aircraft.		AERONAUTICAL MOBILE.
15100–15450	BROADCASTING.	International broadcasting.		International broadcasting.

Worldwide		Region 2		United States	
Band (kHz)	Service	Band(MHz)	Service	Band (MHz)	Allocation
1	2	3	4	5	6
15450-16460	FIXED.				
16460-17360	MARITIME MOBILE. (213)				

Federal Communication Commissions				
Band(KHz) 7	Service 8	Class of station 9	Frequency (MHz) 10	Nature of SERVICES of stations 11
15450–16460	FIXED.	Fixed.		AERONAUTICAL FIXED. INTERNATIONAL FIXED PUBLIC.
16460–16587.1	MARITIME MOBILE.	Ship.		Ship (telephony, duplex).
16587.1–16596.4 (US82)	MARITIME MOBILE.	Ship. Coast.		Ship. Coast (telephony, simplex).
16596.4–16636.5	MARITIME MOBILE.	Ship.		Ship (wide-band telegraphy, facsimile and special transmission systems).
16636.5–16640	MARITIME MOBILE.	Ship. Buoy. Interrogating Coast.		Ship. Buoy. Interrogating Coast (oceanographic data transmission).
16640–16660	MARITIME MOBILE.	Ship.		Ship (wide-band telegraphy, facsimile and special transmission systems).
16660–16694.75	MARITIME MOBILE.	Ship.		Ship (paired, narrow-band direct-printing telegraph and data transmission systems, at speeds not exceeding 100 bauds).
16694.75–16705.8	MARITIME MOBILE.	Ship.		Ship (non-paired, narrow-band direct-printing telegraph and data transmission systems, at speeds not exceeding 100 bauds).
16705.8–16719.8	MARITIME MOBILE.	Ship.		Ship (working, A1 Morse telegraphy).
16719.8–16748.8	MARITIME MOBILE.	Ship.		Ship (calling, A1 Morse telegraphy).
16748.8–16752	MARITIME MOBILE.	Ship.	16749.9 16750.4	Ship (digital selective calling).
16752–16859.4	MARITIME MOBILE.	Ship.		Ship (working, A1 Morse telegraphy).

17360-17700	FIXED.					
17700-17900	BROADCASTING.					
17900-17970	AERONAUTICAL MOBILE. (R)					
17970-18030	AERONAUTICAL MOBILE. (OR)					
18030-18052	FIXED.					
18052-18068	FIXED. Space research.					
18068-19990	FIXED.					
19990-20010	STANDARD FREQUENCY. (201A) (203A) (220)					
20010-21000	FIXED.					
21000-21450	AMATEUR. AMATEUR-SATELLITE.					
21450-21750	BROADCASTING.					
21750-21850	FIXED.					
21850-21870	RADIO ASTRONOMY. (221B)					
21870-22000	AERONAUTICAL FIXED. AERONAUTICAL MOBILE. (R)					

16859.4–17196.9	MARITIME MOBILE.	Coast.		Coast (wide-band and A1 Morse telegraphy, facsimile, special and data transmission systems and direct-printing telegraphy systems).
17196.9–17231.75	MARITIME MOBILE.	Coast.		Coast (paired, narrow-band direct-printing telegraph and data transmission systems, at speeds not exceeding 100 bauds).
17231.75–17232.9	MARITIME MOBILE.	Coast.	17232 17232.5	Coast (digital selective calling).
17232.9–17360	MARITIME MOBILE.	Coast.		Coast (telephony, duplex).
17360–17700	FIXED.	Fixed.		AERONAUTICAL FIXED. INTERNATIONAL FIXED PUBLIC.
17700–17900	BROADCASTING.	International broadcasting.		International broadcasting.
17900–17970 (NG61)	AERONAUTICAL MOBILE. (R)	Aeronautical. Aircraft.		AERONAUTICAL MOBILE.
17970–18030	AERONAUTICAL MOBILE. (OR)	Aeronautical. Aircraft.		AERONAUTICAL MOBILE.
18030–19990	FIXED.	Fixed.		AERONAUTICAL FIXED. INTERNATIONAL FIXED PUBLIC.
19990–20010 (US204)	STANDARD FREQUENCY.	Standard frequency.	20000	Standard frequency.
20010–21000	FIXED.	Fixed.		AERONAUTICAL FIXED. INTERNATIONAL FIXED PUBLIC.
21000–21450	AMATEUR. AMATEUR-SATELLITE.	Amateur. Earth. Space.		AMATEUR. AMATEUR-SATELLITE.
21450–21750	BROADCASTING.	International broadcasting.		International broadcasting.
21750–21850	FIXED.	Fixed.		AERONAUTICAL FIXED. INTERNATIONAL FIXED PUBLIC.
21850–21870 (US74)	RADIO ASTRONOMY.	Radio astronomy.		RADIO ASTRONOMY.
21870–22000	AERONAUTICAL FIXED. AERONAUTICAL MOBILE. (R)	Aeronautical. Aeronautical fixed. Aircraft.		AERONAUTICAL FIXED- AERONAUTICAL MOBILE.

Worldwide		Region 2		United States	
Band (kHZ)	Service	Band (MHz)	Service	Band (MHz)	Allocation
1	2	3	4	5	6
22000-22720	MARITIME MOBILE.				

Band (KHz)	Service	Class of station	Frequency (MHz)	Nature of SERVICES of stations
7	8	9	10	11
22000–22124	MARITIME MOBILE.	Ship.		Ship (telephony, duplex).
22124–22139.5 (US82)	MARITIME MOBILE.	Ship. Coast.		Ship. Coast (telephony, simplex).
22139.5–22160.5	MARITIME MOBILE.	Ship.		Ship (wide-band telegraphy, facsimile and special transmission system).
22160.5–22164	MARITIME MOBILE.	Ship. Buoy. Interrogating Coast.		Ship. Buoy. Interrogating Coast (oceanographic data transmission).
22164–22192	MARITIME MOBILE.	Ship.		Ship (wide-band telegraphy, facsimile and special transmission systems).
22192–22225.75	MARITIME MOBILE.	Ship.		Ship (paired, narrow-band direct-printing telegraph and data transmission systems, at speeds not exceeding 100 bauds).
22225.75–22227	MARITIME MOBILE.	Ship.		Ship (non-paired, narrow-band direct-printing telegraph and data transmission systems, at speeds not exceeding 100 bauds).
22227–22247	MARITIME MOBILE.	Ship.		Ship (calling, A1 Morse telegraphy).
22247–22250	MARITIME MOBILE.	Ship.	22248 22248.5	Ship (digital selective calling).
22250–22310.5	MARITIME MOBILE.	Ship.		Ship (working, A1 Morse telegraphy).
22310.5–22561	MARITIME MOBILE.	Coast.		Coast (wide-band and A1 Morse telegraphy, facsimile, special and data transmission systems and direct-printing telegraphy systems).

22720–23200	FIXED.				
23200–23350	AERONAUTICAL FIXED. AERONAUTICAL MOBILE. (OR)				
23350–24990 (MHz)	FIXED. LAND MOBILE. (222) (222A)				
24.99–25.01	STANDARD FREQUENCY. (203A) (223)			24.99–25.01	G, NG.
25.01–25.07	FIXED. MOBILE (except aeronautical mobile.)			25.01–25.33	NG.
25.07–25.11	MARITIME MOBILE. (224)				
25.11–25.6	FIXED. MOBILE (except aeronautical mobile.)				
25.6–26.1	BROADCASTING.				
26.1–27.5 (225) (226)	FIXED. MOBILE (except aeronautical mobile.)				
				25.33–25.6	G
				25.6–26.1	G, NG.
				26.1–26.48	NG.

22561–22594.75	MARITIME MOBILE.	Coast.		Coast (paired, narrow-band direct-printing telegraph and data transmission systems, at speeds not exceeding 100 bauds).
22594.75–22596	MARITIME MOBILE.	Coast.	22595 22595.5	Coast (digital selective calling).
22596–22720	MARITIME MOBILE.	Coast.		Coast (telephony, duplex).
22720–23200	FIXED.	Fixed.		AERONAUTICAL FIXED. INTERNATIONAL FIXED PUBLIC.
23200–23350	AERONAUTICAL FIXED. AERONAUTICAL MOBILE. (OR)	Aeronautical. Aeronautical fixed. Aircraft.		AERONAUTICAL FIXED. AERONAUTICAL MOBILE.
23350–24990	FIXED.	Fixed.		AERONAUTICAL FIXED. INTERNATIONAL FIXED PUBLIC.
24.99–25.01	STANDARD FREQUENCY.	Standard frequency.	25.0	Standard frequency.
25.01–25.07	LAND MOBILE. (NG 112)	Base. Land mobile.	25.02–25.06 (NG32)	INDUSTRIAL.
25.07–25.76	MARITIME MOBILE.(NG112)	Ship.		Ship (calling, Al Morse telegraphy).
25.76–25.0901	MARITIME MOBILE.	Ship.		Ship (non-paired, narrow-band direct-printing telegraph and data transmission systems, at speeds not exceeding 100 bauds).
25.0901–25.11	MARITIME MOBILE.	Ship.		Ship (working, Al Morse telegraphy).
25.11–25.33	LAND MOBILE.	Base. Land mobile.	25.12–25.32 (NG32)	INDUSTRIAL.
25.6–26.1 (US25)	BROADCASTING.	International broadcasting.		International broadcasting.
26.1–26.48	LAND MOBILE.	Base. Land mobile.	26.11–26.47 (NG32)	Remote pickup broadcast base; remote pickup broadcast mobile.

Worldwide		Region 2		United States	
Band (MHz)	Service	Band (MHz)	Service	Band (MHz)	Allocation
1	2	3	4	5	6
				26.48–26.95 (US10)	G.
27.5–28		27.5–28	METEOROLOGICAL AIDS. FIXED. MOBILE.	27.54–28	G.
28.0–29.7	AMATEUR. AMATEUR-SATELLITE.			28.0–29.7	AMATEUR. AMATEUR-SATELLITE. (US1)
29.7–30.005	FIXED. (228) (229) (231) (232) MOBILE.			29.7–29.89	NG.
				29.89–29.91 (232)	G.
				29.91–30	NG.

	Federal Communications Commission			
Band (MHz)	Service	Class of station	Frequency (MHz)	Nature of SERVICES of stations
7	8	9	10	11
			26.62	Civil air patrol and civil air patrol mobile.
26.95–26.96 (225)	FIXED.	Fixed.	26.955	INTERNATIONAL FIXED PUBLIC.
26.96–27.23 (225) (US1)	PERSONAL.	Fixed. Land mobile.	27.12	Industrial, scientific and medical equipment.
27.23–27.28 (225)	PERSONAL. FIXED. MOBILE.	Fixed. Land mobile.		PERSONAL. PUBLIC SAFETY. INDUSTRIAL. LAND TRANSPORTATION.
27.28–27.41	PERSONAL. LAND MOBILE.	Base. Land mobile.		PERSONAL. INDUSTRIAL.
27.41–27.54	LAND MOBILE.	Base. Land mobile.		INDUSTRIAL.
28.0–29.7	AMATEUR. AMATEUR-SATELLITE.	Amateur. Earth. Space.		AMATEUR. AMATEUR-SATELLITE.
29.7–29.8	LAND MOBILE.	Base. Land mobile.		INDUSTRIAL.
29.8–29.89	FIXED. (232)	Fixed.	29.81–29.88 (NG31)	AERONAUTICAL FIXED. INTERNATIONAL FIXED PUBLIC.
29.91–30	FIXED. (232)	Fixed.	29.92–29.99 (NG31)	AERONAUTICAL FIXED. INTERNATIONAL FIXED PUBLIC.

30.005–30.01	FIXED. (228) (229) (231) MOBILE. SPACE OPERATION (Satellite identification). SPACE RESEARCH.
30.01–37.75	FIXED. (228) (229) (230) (231) MOBILE. (233A)
37.75–38.25	FIXED. (228) (229) (231) MOBILE. Radio astronomy. (233B)

30–30.56	G. (US94)
30.56–32	NG. (NG124)
32–33 (231)	G.
33–34	NG. (NG124)
34–35 (231)	G.
35–36	NG.
36–37 (US220) (231)	G. (US220)
37–38	NG. (NG124)

30.56–32	LAND MOBILE.	Base. Land mobile.		INDUSTRIAL. LAND TRANS-PORTATION. PUBLIC SAFETY.
33–33.01	LAND MOBILE.	Base. Land mobile.		LAND TRANS-PORTATION.
33.01–33.11	LAND MOBILE.	Base. Land mobile.		PUBLIC SAFETY.
33.11–33.41	LAND MOBILE.	Base. Land mobile.		INDUSTRIAL.
33.41–34	LAND MOBILE.	Base. Land mobile.		PUBLIC SAFETY.
35–35.19	LAND MOBILE.	Base. Land mobile.		INDUSTRIAL.
35.19–35.69	LAND MOBILE.	Base. Land mobile.		DOMESTIC PUBLIC. INDUSTRIAL. PUBLIC SAFETY.
35.69–36	LAND MOBILE.	Base. Land mobile.		INDUSTRIAL.
37–37.01	LAND MOBILE.	Base. Land mobile.		INDUSTRIAL.
37.01–37.43	LAND MOBILE.	Base. Land mobile.		PUBLIC SAFETY.
37.43–37.89	LAND MOBILE. (NG59)	Base. Land mobile.		INDUSTRIAL.
37.89–38	LAND MOBILE.	Base. Land mobile.		PUBLIC SAFETY.

Worldwide		Region 2		United States	
Band (MHz)	Service	Band (MHz)	Service	Band (MHz)	Allocation
1	2	3	4	5	6
38.25–41 (235) (236) (236A)	FIXED. (228) (229) (230) (231) MOBILE.			38–39	G. (US81)
				39–40	NG. (US94)
				40–42 (236)	G. (US94) (US210) (US220)
41–50		41–50	FIXED. (228) (231) (237) MOBILE. (233A) (236A)	42–46.6	NG. (NG124)
				46.6–47	G.
				47–49.6	NG. (NG124)
				49.6–50	G.
50–54		50–54	AMATEUR.	50–54	AMATEUR. (US1)
54–68		54–68	FIXED. (228) (237) MOBILE. BROADCASTING.	54–72	NG.
68–73.0		68–73.0	FIXED. MOBILE. BROADCASTING.		
				72–73 (US20)	NG

Federal Communications Commission				
Band (MHz)	Service	Class of station	Frequency (MHz)	Nature of SERVICES of stations
7	8	9	10	11
39–40	LAND MOBILE.	Base. Land mobile.		PUBLIC SAFETY.
			40.68	Industrial, scientific and medical equipment.
42–42.95	LAND MOBILE.	Base. Land mobile.		PUBLIC SAFETY.
42.95–43.19	LAND MOBILE.	Base. Land mobile.		INDUSTRIAL.
43.19–43.69	LAND MOBILE.	Base. Land mobile.		DOMESTIC PUBLIC. INDUSTRIAL. PUBLIC SAFETY.
43.69–44.61	LAND MOBILE.	Base. Land mobile.		LAND TRANSPORTATION.
44.61–46.6	LAND MOBILE.	Base. Land mobile.		PUBLIC SAFETY.
47–47.43	LAND MOBILE.	Base. Land mobile.		PUBLIC SAFETY.
47.43–47.69	LAND MOBILE.	Base. Land mobile.		PUBLIC SAFETY. INDUSTRIAL.
47.69–49.6	LAND MOBILE.	Base. Land mobile.		INDUSTRIAL.
50–54	AMATEUR.	Amateur.		AMATEUR.
54–72	BROADCASTING.	Television broadcasting.	55.25 59.75 61.25 65.75 67.25 71.75	Video, Sound: Channel 2. Video, Sound: Channel 3. Video, Sound: Channel 4.
72–73 (NG56)	FIXED. (NG1) (NG3) (NG49)	Operational fixed.	72.02–72.98 (NG32)	Operational fixed.

73–74.6		73–74.6	RADIO ASTRONOMY. (253A) (253B)	73–74.6	G, NG. (US21) (US100)
74.6–75.4		74.6–75.4 (259)	AERONAUTICAL RADIONAVIGATION.	74.6–75.4	G, NG.
75.4–88		75.4–88	FIXED. MOBILE. BROADCASTING.	75.4–76 (US20)	NG.
				76–88 (US23)	NG.
88–100		88–100	BROADCASTING.	88–108 (US23) (US93)	NG.
100–108		100–108	BROADCASTING.		
108–117.975	AERONAUTICAL RADIONAVIGATION.			108–117.975 (US93)	G, NG.

73-74.6	RADIO ASTRONOMY. (US74)	Radio astronomy.		RADIO ASTRONOMY.
74.6-75.4	AERONAUTICAL RADIONAVIGATION.	Aeronautical radionavigation.	75	Marker beacon.
75.4-76 (NG56)	FIXED. (NG1) (NG3) (NG49)	Operational fixed.	75.42-75.98 (NG32)	Operational fixed.
76-88 (NG21)	BROADCASTING.	Television broadcasting.	77.25 81.75 83.25 87.75	Video } Channel 5. Sound } Video } Channel 6. Sound }
88-108 (NG21)	BROADCASTING.	FM broadcasting. (NG2).	88.1-107.9 (NG36)	FM Channel 201-FM Channel 300.
108-117.975	AERONAUTICAL RADIONAVIGATION.	Radionavigation land.	108.05	Omnidirectional range (VOR).
			108.10	Localizer.
			108.15	Do.
			108.20	Omnidirectional range (VOR).
			108.25	Do.
			108.30	Localizer.
			108.35	Do.
			108.40	Omnidirectional range (VOR).
			108.45	Do.
			108.50	Localizer.
			108.55	Do.
			108.60	Omnidirectional range (VOR).
			108.65	Do.
			108.70	Localizer.
			108.75	Do.
			108.80	Omnidirectional range (VOR).
			108.85	Do.
			108.90	Localizer.
			108.95	Do.
			109.00	Omnidirectional range (VOR).
			109.05	Do.
			109.10	Localizer.
			109.15	Do.
			109.20	Omnidirectional range (VOR).
			109.25	Do.
			109.30	Localizer.
			109.35	Do.
			109.40	Omnidirectional range (VOR).
			109.45	Do.
			109.50	Localizer.
			109.55	Do.
			109.60	Omnidirectional range (VOR).
			109.65	Do.
			109.70	Localizer.
			109.75	Do.
			109.80	Omnidirectional range (VOR).
			109.85	Do.
			109.90	Localizer.
			109.95	Do.

Worldwide		Region 2		United States	
Band (MHz)	Service	Band (MHz)	Service	Band (MHz)	Allocation
1	2	3	4	5	6

Federal Communications Commission				
Band (MHz)	Service	Class of station	Frequency (MHz)	Nature of SERVICES of stations
7	8	9	10	11
			110.00	Omnidirectional range (VOR).
			110.05	Do.
			110.10	Localizer.
			110.15	Do.
			110.20	Omnidirectional range (VOR).
			110.25	Do.
			110.30	Localizer.
			110.35	Do.
			110.40	Omnidirectional range (VOR).
			110.45	Do.
			110.50	Localizer.
			110.55	Do.
			110.60	Omnidirectional range (VOR).
			110.65	Do.
			110.70	Localizer.
			110.75	Do.
			110.80	Omnidirectional range (VOR).
			110.85	Do.
			110.90	Localizer.
			110.95	Do.
			111.00	Omnidirectional range (VOR).
			111.05	Do.
			111.10	Localizer.
			111.15	Do.
			111.20	Omnidirectional range (VOR).
			111.25	Do.
			111.30	Localizer.
			111.35	Do.
			111.40	Omnidirectional range (VOR).
			111.45	Do.
			111.50	Localizer.
			111.55	Do.
			111.60	Omnidirectional range (VOR).
			111.65	Do.
			111.70	Localizer.
			111.75	Do.
			111.80	Omnidirectional range (VOR).
			111.85	Do.
			111.90	Localizer.
			111.95	Do.
			112.00-117.95 (NG34).	Omnidirectional range (VOR).

117.975–132	AERONAUTICAL MOBILE. (R) (201A) (273) (273A)			117.975–121.9375 (273) (US 26) (US 28) (US 85)	GNG.
				121.9375–123.0875 (US30) (US31) (US33) (US80) (US85) (US102) (US213)	NG.
				123.0875–123.5875 (US32) (US33) (US112)	G, NG.
				123.5875–128.8125 (US26) (US85)	G, NG.
				128.8125–132.0125 (US85)	NG.
132–136	AERONAUTICAL MOBILE. (R) (273A) (274A) (274B)			132.0125–136 (US26) (US85)	G, NG.
136–137	SPACE RESEARCH (Space-to-earth). (281A) (281AA)			136–137	G, NG.

117.975–121.9375	AERONAUTICAL MOBILE. (R)	Aeronautical. Aircraft.	118.000–121.400 (NG67)	Airdrome conrtol.
			121.5	Aeronautical Mobile. Maritime Mobile. Maritime Radiodetermination. Ship. Survival craft and equipment.
			121.600–121.925	Emergency position indicating radiobeacon. Emergency locator transmitter.
			(NG67)	Aeronautical utility mobile.
			121.950 (NG67)	Aeronautical utility land. ELT test.
121.9375–123.0875	AERONAUTICAL MOBILE.	Aeronautical. Aircraft.	121.950 (NG67)	Aviation instructional.
			121.975–123.075 (NG67)	Private aircraft.
123.0875–123.5875 (NG67)			123.100	Aeronautical search and rescue. Ship. Survival craft and equipment.
			123.125–123.275	Flight test.
			123.300	Aviation instructional.
			123.325–123.475	Flight test.
			123.500	Aviation instructional.
			123.525–123.575	Flight test.
123.5875–128.8125	AERONAUTICAL MOBILE. (R)	Aeronautical. Aircraft.	123.600–128.800 (NG67)	AERONAUTICAL MOBILE.
128.8125–132.0125	AERONAUTICAL MOBILE. (R)	Aeronautical. Aircraft.	128.825–132.000 (NG67)	AERONAUTICAL MOBILE.
132.0125–136	AERONAUTICAL MOBILE. (R)	Aeronautical. Aircraft.	132.025–135.975 (NG67)	AERONAUTICAL MOBILE.
136–137	SPACE RESEARCH.	Space.		Space.

§ 2.106 Table of Frequency Allocations—Continued

Worldwide		Region 2		United States	
Band (MHz)	Service	Band (MHz)	Service	Band (MHz)	Allocation
1	2	3	4	5	6
137–138	METEOROLOGICAL-SATELLITE. SPACE OPERATION (Telemetering and tracking). SPACE RESEARCH (Space-to-earth). (275A) (281E)			137–138	G, NG.
138–143.6		138–143.6	FIXED. MOBILE. *Radiolocation.* Space research (Space-to-earth). (283A).	138–144 (US10)	G.
143.6–143.65		143.6–143.65	FIXED. MOBILE. SPACE RESEARCH (Space-to-earth). *Radiolocation.* (283A)		
143.65–144		143.65–144	FIXED. MOBILE. *Radiolocation.* Space research (Space-to-earth).		
144–146	AMATEUR. AMATEUR-SATELLITE.			144–146	AMATEUR. AMATEUR-SATELLITE. (US1)
146–148		146–148	AMATEUR.	146–148	AMATEUR. (US1)
148–149.9 (285A)		148–149.9 (285A)	FIXED. MOBILE.	148–149.9 (285A) (US10)	G.
149.9–150.05 (285B) (285C)	RADIONAVIGATION-SATELLITE.			149.9–150.05	G, NG. (US100)
150.05–174		150.05–174	FIXED. MOBILE.	150.05–150.8	G. (US216)

Band (MHz) 7	Service 8	Class of station 9	Frequency (MHz) 10	Nature {OF SERVICES of stations 11
137-138	METEOROLOGICAL-SATELLITE. SPACE OPERATION. SPACE RESEARCH.	Space.		Space.
				• •
144-146	AMATEUR. AMATEUR-SATELLITE.	Amateur. Earth. Space.		AMATEUR. AMATEUR-SATELLITE.
146-148	AMATEUR.	Amateur.		AMATEUR.
			148.15	Civil air patrol land. Civil air patrol mobile.
149.9-150.05	RADIONAVIGATION-SATELLITE.	Space.		RADIONAVIGATION-SATELLITE.

				150.8–157.0375	NG. (US216)
					(NG124)
					(US77)
(201A)					
(287)					
					(US106) US107)

150.8-150.98	LAND MOBILE. (NG112)	Base. Land mobile.		LAND TRANSPORTATION. (NG51).
150.980 151.4825	LAND MOBILE. (NG112)	Base. Land mobile.		PUBLIC SAFETY (NG51).
151.4825–151.4975	LAND MOBILE.	Base. Land mobile.		INDUSTRIAL & PUBLIC SAFETY (NG51).
151.4975–152	LAND MOBILE.	Base. Land mobile.		INDUSTRIAL.
152–152.255	LAND MOBILE.	Base. Land mobile.		DOMESTIC-PUBLIC.
152.255–152.465	LAND MOBILE.	Base. Land mobile.		LAND TRANSPORTATION. (NG38)(NG116)
152.465–152.495	LAND MOBILE.	Base. Land mobile.		INDUSTRIAL. (NG116)
152.495–152.855	LAND MOBILE.	Base. Land mobile.		DOMESTIC PUBLIC. (NG4)
152.855–153.7325	LAND MOBILE.	Base. Land mobile.		INDUSTRIAL. (NG4)
153.7325–154.4825	LAND MOBILE.	Base. Land mobile. Fixed.	154.2	INDUSTRIAL, PUBLIC SAFETY (NG26)(NG57) (Earth Telecommand).
154.4825–154.6375	LAND MOBILE. (NG112)	Base. Land mobile.		Industrial.
154.6375–156.25	LAND MOBILE. (NG119)			PUBLIC SAFETY.
156.250–157.0375	MARITIME MOBILE. (NG5)	----------------------	156.250 156.275 156.300 156.325 156.350 156.375 156.400 156.425 156.450 156.475 156.500 156.525 156.550 156.575 156.600 156.625 156.650 156.675 156.700 156.725 156.75	MARITIME MOBILE. (NG117) Do. Do. MARITIME MOBILE. (NG24) MARITIME MOBILE. Do. MARITIME MOBILE. (NG24) MARITIME MOBILE. Do. MARITIME MOBILE. (NG24) MARITIME MOBILE. Do. Do. Do. Do. Do. Do. MARITIME MOBILE. MARITIME RADIODETERMINATION. Ship. Emergency position indicating radiobeacon.

§ 2,106 Table of Frequency Allocations—Continued

Worldwide		Region 2		United States	
Band (MHz)	Service	Band (MHz)	Service	Band (MHz)	Allocation
1	2	3	4	5	6
				157.0375–157.1875 (US214)	G.
(287A)		(287A)		157.1875–162.0125	NG. (US77) (US200) (NG111) (US 223) (NG124)

Band (MHz)	Service	Class of station	Frequency (MHz)	Nature of SERVICES of stations
7	8	9	10	11
			156.8	MARITIME MOBILE (distress, safety and calling). MARITIME RADIODETERMINATION. Ship. Emergency position indicating radiobeacon.
			156.850	MARITIME MOBILE.
			156.875	Do.
			156.900	Do.
			156.925	Do.
			156.950	Do.
			156.975	Do.
			157.100	Do.
			157.025	Do.
			157.100	US Coast Guard Liaison Frequency.
157.1875–157.450	MARITIME MOBILE. (NG5)	Ship.	157.200	MARITIME MOBILE.
			157.225	Do.
			157.250	Do.
			157.275	Do.
			157.300	Do.
			157.325	Do.
			157.350	Do.
			157.375	Do.
			157.400	Do.
			157.425	Do.
157.45–157.725	LAND MOBILE.	Base. Land mobile.		LAND TRANSPORTATION. (NG5) (NG38) (NG116)
157.725–157.755	LAND MOBILE.	Base. Land mobile.		INDUSTRIAL. (NG116)
157.755–158.115	LAND MOBILE.	Base. Land mobile.		DOMESTIC PUBLIC.
158.115–158.475	LAND MOBILE. (NG112)	Base. Land mobile.		INDUSTRIAL.
158.475–158.715	LAND MOBILE.	Base. Land mobile.		DOMESTIC PUBLIC.
158.715 159.480	LAND MOBILE. (NG112)	Base. Land mobile.		PUBLIC SAFETY. (NG70)
159.480 161.575	LAND MOBILE. (NG112)	Base. Land mobile.		LAND TRANSPORTATION. (NG6) (NG26) (NG28) (NG70)

			(233A)	162.0125-173.2	G. (US8) (US11) (US13) (US216) (US223)
				173.2-173.4	N.G
				173.4-174.0	G.
174-216		174-216 (294)	FIXED. MOBILE. BROADCASTING.	174-216	NG. NG115
216-220		216-220	FIXED. MOBILE. RADIOLOCATION.	216-220	G, NG (US114) (US210) NG121
220-225		220-225	AMATEUR RADIOLOCATION.	220-225	G, NG. (US34)

161.575–161.625 (US77)	MARITIME MOBILE.	Coast.	161.6	MARITIME MOBILE (NG6) (NG17) (NG26)
161.625 161.775	LAND MOBILE.	Base. Land mobile.		Remote pickup broadcast base; remote pickup broadcast mobile. (NG6) (NG26)
161.775–162.0125	MARITIME MOBILE. (NG5)	Coast.	161.800 161.825 161.850 161.875 161.900 161.925 161.950 161.975 162.000	Coast (NG26). Do. Do. Do. Coast. Do. Do. Do. Do.
			166.25 170.15	PUBLIC SAFETY. Remote pickup. Do.
			170.425 170.475 170.575 171.425 171.475 171.575 172.225 172.275 172.375	PUBLIC SAFETY. Do. Do. Do. Do. Do. Do. Do. Do.
173.2–173.4	Fixed. Land mobile.	Base. Fixed mobile.		PUBLIC SAFETY. INDUSTRIAL.
174–216	BROADCASTING.	Television broadcasting.	175.25 179.75 181.25 185.75 187.25 191.75 193.25 197.75 199.25 203.75 205.25 209.75 211.25 215.75	Video } Channel 7. Sound } Video } Channel 8. Sound } Video } Channel 9. Sound } Video } Channel 10. Sound } Video } Channel 11. Sound } Video } Channel 12. Sound } Video } Channel 13. Sound }
216–220	MARITIME MOBILE. Land mobile.	Coast. Ship. Telemetering land. Telemetering mobile.		MARITIME MOBILE INDUSTRIAL
220–225	Amateur.	Amateur.		AMATEUR.

Worldwide		Region 2		United States	
Band (MHz) 1	Service 2	Band (MHz) 3	Service 4	Band (MHz) 5	Allocation 6
225–235		225–235	FIXED. MOBILE.	225–328.6 (308A) (310) (US98)	G.
235–267	FIXED. MOBILE. (201A) (305A) (308A)				
267–272	FIXED, MOBILE. Space operation Telemetering). (308A) (309A) (309B)				
272–273	FIXED. MOBILE. SPACE OPERATION (Telemetering). (308A) (309A)				
273–328.6	FIXED. MOBILE. (308A) (310) (310A)				
328.6–335.4	AERONAUTICAL RADIONAVIGATION. (311)			328.6–335.4	G, NG.
335.4–399.9	FIXED. MOBILE. (308A)			335.4–399.9	G. (308A)
399.9–400.05	RADIONAVIGATION-SATELLITE. (311A) (285C)			399.9–400.05	G, NG. (US100)
400.05–400.15	STANDARD FREQUENCY-SATELLITE. (312B) (313) (314)			400.05–400.15	G. (312B)
400.15–401	METEOROLOGICAL AIDS. METEOROLOGICAL-SATELLITE (Maintenance telemetering). SPACE RESEARCH (Telemetering and tracking).			400.15–401	G, NG.

Band (MHz) 7	Service 8	Class of station 9	Frequency (MHz) 10	Nature of services of stations 11
			243	Survival craft and equipment. Emergency position indicating radiobeacon. Emergency locator transmitter.
328.6–335.4	AERONAUTICAL RADIONAVIGATION. (311)	Radionavigation land.		Glide path.
399.9–400.05	RADIONAVIGATION-SATELLITE.	Space.		RADIONAVIGATION-SATELLITE.
400.05–400.15	STANDARD FREQUENCY-SATELLITE.	Space.	400.1	Standard frequency.
400.15–401	METEOROLOGICAL AIDS. (US70). SPACE RESEARCH (Telemetering and tracking).	Radiosonde. Space.		Radiosonde. Space.

401–402 (314) (315C)	METEOROLOGICAL AIDS. SPACE OPERATION (Telemetering). (315A) Fixed. Meteorological-Satellite (Earth-to-space). Mobile except aeronautical mobile.			401–402	G, NG. (315C)
402–403 (315C)	METEOROLOGICAL AIDS. Fixed. Meteorological-Satellite (Earth-to-space). Mobile except aeronautical mobile.			402–403	G, NG. (315C)
403–406 (315)	METEOROLOGICAL AIDS. Fixed. Mobile except aeronautical mobile.			403–406	G, NG.
406–406.1	MOBILE-SATELLITE (Earth-to-space). (317A)(317B)			406–406.1	G, NG. (317A)
406.1–410 (233B)	FIXED. MOBILE except aeronautical mobile. RADIO ASTRONOMY.			406.1–410	G, NG. (US13) (US74) (US117)
410–420	FIXED. MOBILE except aeronautical mobile.			410–420	G. (US13)
420–450 (320A)		420–450	RADIOLOCATION. Amateur. (318) (319A) (319B) (320A)	420–450	G, NG (US217) (US87) (US7) (320A) (US35)
450–460 (318) (318B) (318C) (319A)	FIXED. MOBILE.			450–470	NG (US87) (US100) (US201) (US209) (US216) (NG124)

401–402	METEOROLOGICAL AIDS. (US70) SPACE OPERATION (Telemetering). (315A) Meteorological-Satellite (Earth-to-space).	Earth. Radiosonde. Space.		Earth. Radiosonde. Space.
402–403	METEOROLOGICAL AIDS. (US70) Meteorological-Satellite (Earth-to-space).	Earth. Radiosonde.		Earth. Radiosonde.
403–406	METEOROLOGICAL AIDS. (US70)	Radiosonde.		Radiosonde.
406–406.1	MOBILE-SATELLITE (Earth-to-space).	Earth.		Emergency position-indicating radiobeacon.
406.1–410	RADIO ASTRONOMY.	Radio astronomy.		RADIO ASTRONOMY.
420–450	Amateur. Amateur-Satellite.	Amateur. Earth. Space.		AMATEUR. AMATEUR-SATELLITE.
450–451	LAND MOBILE.	Base. Land mobile.		Remote pickup broadcast base; remote pickup broadcast mobile.
451–454	LAND MOBILE. (NG 112)	Base. Land mobile.		PUBLIC SAFETY. INDUSTRIAL. LAND TRANSPORTATION.
454–455	LAND MOBILE. (NG112)	Base. Land mobile.		DOMESTIC PUBLIC. (NG12)
455–456	LAND MOBILE.	Base. Land mobile.		Remote pickup broadcast base; remote pickup broadcast mobile.
456–459	LAND MOBILE. (NG112)	Land mobile.		PUBLIC SAFETY. INDUSTRIAL. LAND TRANSPORTATION.

§ 2.106 Table of Frequency Allocations—Continued

Worldwide		Region 2		United States	
Band (MHz) 1	Service 2	Band (MHz) 3	Service 4	Band (MHz) 5	Allocation 6
460–470 (324B)	FIXED. MOBILE. Meteorological-Satellite (Space-to-earth). (318A)				(US100) (US201) (US209) (NG 124)
470–890 (332A)		470–890 (329A) (332) (332A)	BROADCASTING.	470–902	NG. (NG30) (NG43) (NG63) (US36) (US88) (US100) (US116) (US215)

Federal Communications Commission				
Band (MHz)	Service	Class of station	Frequency (MHz)	Nature of services of stations
7	8	9	10	11
459–460	LAND MOBILE. (NG112)	Base. Land mobile.		DOMESTIC PUBLIC. (NG12)
460–462.5375	LAND MOBILE.	Base. Land mobile.		PUBLIC SAFETY. INDUSTRIAL. LAND TRANSPORTATION.
462.5375–462.7375	LAND MOBILE.	Base. Land mobile.		PERSONAL.
462.7375–465.0125	LAND MOBILE.	Base. Land mobile.		PUBLIC SAFETY. INDUSTRIAL. LAND TRANSPORTATION.
465.0125–467.5375	LAND MOBILE.	Land mobile.		PUBLIC SAFETY. INDUSTRIAL. LAND TRANSPORTATION.
467.5375–467.7375	LAND MOBILE.	Land mobile.		PERSONAL.
467.7375–470	LAND MOBILE.	Land mobile.		PUBLIC SAFETY. INDUSTRIAL. LAND TRANSPORTATION.
470–512	BROADCASTING. LAND MOBILE. (NG66) (NG114)	Television broadcasting. Land mobile. Base.		BROADCASTING. PUBLIC SAFETY. INDUSTRIAL. LAND TRANSPORTATION. DOMESTIC PUBLIC.
512–806	BROADCASTING.	Television broadcasting.		
806–821	LAND MOBILE.	Land mobile.		Conventional and Trunked Systems.
821–825	LAND MOBILE.			Reserve.
825–845	LAND MOBILE.	Land mobile.		Cellular Systems.
845–851	LAND MOBILE.			Reserve.
851–866	LAND MOBILE.	Base.		Conventional and Trunked Systems.
866–870	LAND MOBILE.			Reserve.
870–890	LAND MOBILE.	Base.		Cellular Systems.

890-942		890-942 (339A) (340)	FIXED. RADIOLOCATION.	902-928	G. (US36) (US115) (US218)
				928-960	NG. (US36) (US116) (US215)
942-960		942-960 (339A)	FIXED.		
960-1215	AERONAUTICAL RADIONAVIGATION. (341)			960-1215	G, NG. (US224)
1215-1300	RADIOLOCATION. Amateur.			1215-1300	G, NG. (US34)
1300-1350	AERONAUTICAL RADIONAVIGATION. (346) Radiolocation.			1300-1350	G, NG.
1350-1400		1350-1400 (349A)	RADIOLOCATION.	1350-1400	G.
1400-1427	RADIO ASTRONOMY.			1400-1427	G, NG.
1427-1429	FIXED. MOBILE except aeronautical mobile. SPACE OPERATION (Telecommand).			1427-1429	G, NG. (US211)
1429-1435		1429-1435	FIXED. MOBILE.	1429-1435	G, NG.

890-902	LAND MOBILE.			Reserve.
			915	Industrial, scientific, and medical equipment.
928-929 (NG120)	Fixed.	Operational fixed.		
929-947	Land Mobile.			Reserve
947-952	Fixed.			
952-960 (NG10) (NG120)	FIXED. Mobile (Multiple address systems)	International fixed public (Puerto Rico and Virgin Islands only). International control. Operational fixed.		
960-1215	AERONAUTICAL RADIONAVIGATION. (341)			
1215-1300	Amateur.	Amateur.		AMATEUR.
1300-1350	AERONAUTICAL RADIONAVIGATION. (346)			
1400-1427	RADIO ASTRONOMY. (US74)			
1427-1429	SPACE OPERATION (Telecommand). Fixed (Telemetering). Land mobile (Telemetering and telecommand).	Base. Earth. Fixed. Land mobile.		INDUSTRIAL. LAND TRANSPORTATION. PUBLIC SAFETY. Earth (telecommand). Base (telecommand). Fixed (telemetering). Land mobile (telemetering).
1429-1435	Fixed (Telemetering). Land mobile (telemetering and telecommand).	Base. Fixed. Land mobile.		INDUSTRIAL. LAND TRANSPORTATION. PUBLIC SAFETY. Base (telecommand). Fixed (telemetering). Land mobile (telemetering).

§ 2.106 Table of Frequency Allocations—Continued

Worldwide		Region 2		United States	
Band (MHz)	Service	Band (MHz)	Service	Band (MHz)	Allocation
1	2	3	4	5	6
1435–1525		1435–1525	MOBILE. Fixed.	1435–1535	G, NG. (US78)
1525–1535		1525–1535	SPACE OPERATION (Telemetering). (350A) Earth exploration-satellite. Fixed. Mobile. (350D)		
1535–1542.5	MARITIME MOBILE-SATELLITE (352E)			1535–1542.5	G, NG. (352E) (US39)
1542.5–1543.5	AERONAUTICAL MOBILE-SATELLITE. (R) MARITIME MOBILE-SATELLITEl (352F)			1542.5–1543.5	G, NG. (352F) (US39)
1543.5–1558.5	AERONAUTICAL MOBILE-SATELLITE. (R) (352G)			1543.5–1558.5	G, NG (352G) (US39)
1558.5–1636.5	AERONAUTICAL RADIONAVIGATION. (352A) (352B) (352K)			1558.5–1636.5	G, NG. (352A) (352B) (US39) (US40) (US208)
1636.5–1644	MARITIME MOBILE-SATELLITE. (352H)			1636.5–1644	G, NG. (352H) (US39)
1644–1645	AERONAUTICAL MOBILE-SATELLITE. (R) MARITIME MOBILE-SATELLITE. (352I)			1644–1645	G, NG. (352I) (US39)
1645–1660	AERONAUTICAL MOBILE SATELLITE. (R) (352J)			1645–1660	G, NG. (352J) (US39)
1660–1670	METEOROLOGICAL AIDS. RADIO ASTRONOMY. (353A) (354A)			1660–1670	G, NG. (US74) (US99) (US100)

Band (MHz)	Service	Class of station	Frequency (MHz)	Nature of SERVICES of stations
7	8	9	10	11
1435–1535	MOBILE.	Aeronautical telemetering.		AVIATION.
1535–1542.5	MARITIME MOBILE-SATELLITE.	Coast. Ship. Space.		MARITIME MOBILE-SATELLITE.
1542.5–1543.5	AERONAUTICAL MOBILE SATELLITE. (R) MARITIME MOBILE-SATELLITE.	Aeronautical. Aircraft. Coast. Ship. Space.		AERONAUTICAL MOBILE-SATELLITE. (R) MARITIME MOBILE-SATELLITE.
1543.5–1558.5	AERONAUTICAL MOBILE SATELLITE. (R)	Aeronautical. Aircraft. Space.		AERONAUTICAL MOBILE-SATELLITE. (R)
1558.5–1592.5	AERONAUTICAL RADIONAVIGATION.	Radionavigation land.		
1592.5–1622.5	AERONAUTICAL RADIONAVIGATION.	Radionavigation land. Radionavigation mobile.		Collision avoidance. (Provisional)
1622.5–1636.5	AERONAUTICAL RADIONAVIGATION.			
1636.5–1644	MARITIME MOBILE-SATELLITE.	Earth. Ship.		MARITIME MOBILE-SATELLITE.
1644–1645	AERONAUTICAL MOBILE-SATELLITE. (R) MARITIME MOBILE-SATELLITE.	Aircraft. Earth. Ship.		AERONAUTICAL MOBILE-SATELLITE. MARITIME MOBILE-SATELLITE.
1645–1660	AERONAUTICAL MOBILE-SATELLITE. (R)	Aircraft. Earth.		AERONAUTICAL MOBILE-SATELLITE. (R)
1660–1670	METEOROLOGICAL AIDS. RADIO ASTRONOMY.	Radio astronomy. Radiosonde.		Radio astronomy. Radiosonde.

1670–1690	FIXED. METEOROLOGICAL AIDS. METEOROLOGICAL SATELLITE. (Space-to-earth) (324A) MOBILE except aeronautical mobile.			1670–1690	G, NG. (324A)
1690–1700		1690–1700	METEOROLOGICAL AIDS. METEOROLOGICAL-SATELLITE (Space-to-earth). (324B) (354A)	1690–1700	G, NG. (324B) (US100)
1700–1710		1700–1710	FIXED. MOBILE. SPACE RESEARCH (Space-to-earth) (354D)	1700–1710	G, NG.
1710–1770		1710–1770	FIXED. MOBILE. (352K) (356A)	1710–1850	G. (US100)
1770–1790 (356AA)		1770–1790	FIXED. MOBILE. Meteorological-Satellite. (356AA) (356A)		
1790–2290 (356AB) (356ABA)		1790–2290	FIXED. MOBILE. (356A) (356AB) (356ABA)	1850–2200	NG. (US90) (US111) (US219) (US222)
				2200–2290	G.
2290–2300		2290–2300	FIXED. MOBILE. SPACE RESEARCH (Space-to-earth).	2290–2300	G, NG.

1670–1690	METEOROLOGICAL AIDS. METEOROLOGICAL-SATELLITE.	Radiosonde. Space.		
1690–1700	METEOROLOGICAL AIDS. METEOROLOGICAL-SATELLITE.	Radiosonde. Space.		
1700–1710	SPACE RESEARCH. METEOROLOGICAL-SATELLITE.	Space.		
1850–1990 (NG8)	FIXED.	International control. Operational fixed.		
1990–2110 (NG118)	FIXED. MOBILE.	Television pickup. Television STL. Television intercity relay.		
2110–2130 (NG10) (NG23)	FIXED.	Domestic fixed public.		
2130–2150 (NG10) (NG23)	FIXED.	Operational fixed. International control.		
2150–2160 (NG23) (NG45)	FIXED.	Fixed.		MULTIPOINT DISTRIBUTION. Operational fixed.
2160–2180 (NG10) (NG23)	FIXED.	Fixed.		MULTIPOINT DISTRIBUTION DOMESTIC FIXED PUBLIC.
2180–2200 (NG10) (NG23)	FIXED.	Operational fixed. International control.		
2290–2300	SPACE RESEARCH	Space.		

§ 2.106 Table of Frequency Allocations—Continued

Worldwide		Region 2		United States	
Band (MHz)	Service	Band (MHz)	Service	Band (MHz)	Allocation
1	2	3	4	5	6
2300–2450 (357)		2300–2450	RADIOLOCATION. Amateur. Fixed. Mobile. (357)	2300–2450	G, NG. (US34)
2450–2500		2450–2500 (357)	FIXED. MOBILE. RADIO-LOCATION.	2450–2500	NG. (US41)
2500–2550 (361A) (361B) (364C)		2500–2535	BROADCASTING-SATELLITE. (361B) FIXED. (364C) FIXED-SATELLITE (Space-to-earth). MOBILE except aeronautical mobile. (361A) (364E)	2500–2535	NG. (NG8) (NG47) (NG101) (NG102) (US205) (US211)
		2535–2550	BROADCASTING-SATELLITE. (361B) FIXED. (364C) MOBILE except aeronautical mobile. (361A)	2535–2655	NG. (NG8) (NG47) (NG101) (US205) (US211)
2550–2655	BROADCASTING-SATELLITE. (361B) FIXED. (364C) MOBILE except aeronautical mobile.				
2655–2690 (361B) (364C) (364D) (364G) (364H)		2655–2690	BROADCASTING-SATELLITE. (361B) (364H) FIXED. (364C) (364D) FIXED-SATELLITE (Earth-to-space). MOBILE except aeronautical mobile. (364E) (364G)	2655–2690	NG. (NG8) (NG47) (NG101) (NG102) (US205) (US211)
2690–2700 (233B) (364A)	RADIO ASTRONOMY.			2690–2700	G, NG. (US74) (US100)
2700–2900 (366)	AERONAUTICAL RADIONAVIGATION. (346). Radiolocation.			2700–2900	G. (346) (366) (US42) (US43)

Band (MHz)	Service	Class of station	Frequency (MHz)	Nature of SERVICES of stations
7	8	9	10	11
2300–2450 (357)	Amateur.	Amateur.		
2450–2500 (357)	FIXED. MOBILE. Radiolocation.			
2500–2535	BROADCASTING-SATELLITE. FIXED. FIXED-SATELLITE.	Instructional television fixed. Operational fixed. Space.		BROADCASTING-SATELLITE. FIXED. FIXED-SATELLITE.
2535–2655	BROADCASTING-SATELLITE. FIXED.	Instructional television fixed. Operational fixed. Space.		BROADCASTING-SATELLITE. FIXED.
2655–2690	BROADCASTING-SATELLITE. FIXED. FIXED-SATELLITE.	Instructional television fixed. Operational fixed. Space. Earth.		BROADCASTING-SATELLITE. FIXED. FIXED SATELLITE.
2690–2700	RADIO ASTRONOMY.	Radio astronomy.		

2900–3100	RADIONAVIGATION. (367) (367A) (367B). Radiolocation.			2900–3100	G, NG.
3100–3300 (354) (368) (369)	RADIOLOCATION.			3100–3300	G, NG. (369) (US 110)
3300–3400		3300–3400	RADIOLOCATION. Amateur.	3300–3500	G, NG. (US108)
3400–3500		3400–3500	FIXED-SATELLITE (Space-to-earth). RADIO-LOCATION. Amateur.		
3500–3700		3500–3700	FIXED. FIXED-SATELLITE (Space-to-earth). MOBILE. RADIO-LOCATION.	3500–3700	G, NG. (US110)
3700–4200		3700–4200	FIXED. FIXED-SATELLITE (Space-to-earth). MOBILE.	3700–4200	NG.
4200–4400 (352A) (379A)	AERONAUTICAL RADIO-NAVIGATION.			4200–4400	G, NG. (US47)
4400–4700	FIXED. FIXED-SATELLITE (Earth-to-space). MOBILE.			4400–4990	G. (US203)
4700–4990 (233B) (382A) (382B)	FIXED. MOBILE.				
4990–5000		4990–5000 (383A)	RADIO ASTRONOMY.	4990–5000	G, NG. (US74) (US100)
5000–5250 (352A) (352B) (383B)	AERONAUTICAL RADIONAVIGATION.			5000–5250	G, NG. (352A) (352B) (383B) (US118) (US211)
5250–5255	RADIOLOCATION. Space research.			5250–5350	G, NG. (US110)
5255–5350	RADIOLOCATION.				
5350–5460	AERONAUTICAL RADIONAVIGATION. (385) Radiolocation.			5350–5460	G, NG.

2900-3100	MARITIME RADIONAVIGATION. Radiolocation. (US44)			
3100-3300	Radiolocation.	Radiolocation land. Radiolocation mobile.		RADIOLOCATION.
3300-3500	Amateur. Radiolocation.	Amateur. Radiolocation land. Radiolocation mobile.		AMATEUR. RADIOLOCATION.
3500-3700	Radiolocation.	Radiolocation land. Radiolocation mobile.		RADIOLOCATION.
3700-4200	FIXED. FIXED-SATELLITE.	Common carrier fixed. Space.		DOMESTIC PUBLIC. (NG41) FIXED-SATELLITE.
4200-4400	AERONAUTICAL RADIO-NAVIGATION.	Altimeter.		
4990-5000	RADIO ASTRONOMY.	Radio astronomy.		
5000-5250	AERONAUTICAL RADIONAVIGATION.			
5250-5350	Radiolocation.	Radiolocation land. Radiolocation mobile.		RADIOLOCATION.
5350-5460	AERONAUTICAL RADIONAVIGATION. (385) Radiolocation. (US48)			

Worldwide		Region 2		United States	
Band (MHz) 1	Service 2	Band (MHx) 3	Service 4	Band (MHz) 5	Allocation 6
5460-5470	RADIONAVIGATION. (385) Radiolocation.			5460-5470	G, NG.
5470-5650 (387)	MARITIME RADIONAVIGATION. Radiolocation.			5470-5600	G, NG.
				5600-5650	G, NG
5650-5670 (388)	RADIOLOCATION. Amateur.			5650-5925	G, NG (391) (US34) (US100)
5670-5725 (388) (389A)	RADIOLOCATION. Amateur. Space research (deep space).				
5725-5925 (391)		5725-5925	RADIOLOCATION. Amateur. (391) (391A)		
5925-6425	FIXED. FIXED-SATELLITE (Earth-to-space). MOBILE.			5925-6425	NG.
6425-7250 (379A) (392AA) (392B) (393)	FIXED. MOBILE.			6425-7125	NG.
				7125-7250	G. (392B)

Federal Communication Commissions

Band (MHz) 7	Service 8	Class of station 9	Frequency (MHz) 10	Nature of services of stations 11
5460–5470	RADIONAVIGATION. (385) (US65) Radiolocation. (US49)			
5470–5600	MARITIME RADIONAVIGATION. (US65) Radiolocation. (US50)			
5600–5650	MARITIME RADIONAVIGATION. (US65) METEOROLOGICAL AIDS. (387) Radiolocation. (US51)			
5650–5925	Amateur.	Amateur.	5800	Industrial, scientific and medical equipment.
5925–6425	FIXED. FIXED-SATELLITE.	Common carrier fixed. Fixed earth.		DOMESTIC PUBLIC. (NG41) FIXED-SATELLITE.
6425–6525	MOBILE.	Common carrier land. Common carrier mobile.		
6525–6575	FIXED.	Operational fixed.		
6575–6625 (NG8)	FIXED.	International control. Operational fixed.		
6625–6875 (NG8)	FIXED. FIXED-SATELLITE. (NG103)	International control. Operational fixed. Space.		
6875–7125 (NG118)	FIXED. FIXED-SATELLITE. (NG103) MOBILE.	Space. Television pickup. Television STL. Television intercity relay.		

7250-7300 (392D) (392G)	FIXED-SATELLITE. (Space-to-earth).			7250-7300	G. (392D) (US100)
7300-7450 (392D)	FIXED. FIXED-SATELLITE (Space-to-earth). MOBILE.			7300-7750	G. (392D)
7450-7550 (392D)	FIXED. FIXED-SATELLITE (Space-to-earth). METEOROLOGICAL-SATELLITE (Space-to-earth). MOBILE.				
7550-7750 (392D)	FIXED. FIXED-SATELLITE (Space-to-earth). MOBILE.				
7750-7900	FIXED. MOBILE.			7750-7900	G.
7900-7975	FIXED. FIXED-SATELLITE (Earth-to-space). MOBILE.			7900-7975	G.
7975-8025 (392H)	FIXED-SATELLITE (Earth-to-space).			7975-8025	G. (US100)
8025-8175		8025-8175	EARTH EXPLORATION-SATELLITE (Space-to-earth). FIXED. FIXED-SATELLITE (Earth-to-space). MOBILE.	8025-8400	G.
8175-8215		8175-8215	EARTH-EXPLORATION-SATELLITE (Space-to-earth). FIXED. FIXED-SATELLITE (Earth-to-space). METEOROLOGICAL SATELLITE (Earth-to-space). MOBILE.		
8215-8400		8215-8400	EARTH EXPLORATION-SATELLITE (Space-to-earth). FIXED. FIXED-SATELLITE (Earth-to-space). MOBILE.		

§ 2.106 Table of Frequency Allocations—Continued

Worldwide		Region 2		United States	
Band (GHz)	Service	Band (GHz)	Service	Band (GHz)	Allocation
1	2	3	4	5	6
8400-8500	FIXED. MOBILE. SPACE RESEARCH (Space-to-earth).			8400-8500	G, NG.
8500-8750	RADIOLOCATION.			8500-9000	G, NG. (US53) (US110)
8750-8850	RADIOLOCATION. AERONAUTICAL RADIONAVIGATION.				
8850-9000	RADIOLOCATION.				
9000-9200	AERONAUTICAL RADIONAVIGATION. Radiolocation.			9000-9200	G, NG.
9200-9300	RADIOLOCATION.			9200-9300	G, NG. (US110)
9300-9500 (399)	RADIONAVIGATION. (367)(367A)(367B) Radiolocation.			9300-9500	G, NG.
9500-9800	RADIOLOCATION.			9500-10000	G, NG. (401A) (US110)
9800-10000 (401A)	RADIOLOCATION. Fixed.				
10000-10500 (401A)	RADIOLOCATION. Amateur.			10000-10500	G, NG. (401A) (US58) (US108)
10.5-10.55		10.5-10.55	RADIOLOCATION. (404).	10.5-10.55	G, NG. (US59)
10.55-10.6	FIXED. MOBILE. Radiolocation.			10.55-10.68	NG.
10.6-10.68	FIXED. MOBILE. RADIO ASTRONOMY. Radiolocation.				

Federal Communications Commission				
Band (GHz)	Service	Class of station	Frequency (GHz)	Nature of SERVICES of stations
7	8	9	10	11
8400-8500	SPACE RESEARCH.	Space.		
8500-9000	Radiolocation.	Radiolocation land. Radiolocation mobile.	8800	RADIOLOCATION. Airborne doppler radar.
9000-9200	AERONAUTICAL RADIONAVIGATION. (346) (US54) Radiolocation. (US48)			
9200-9300	Radiolocation.	Radiolocation land. Radiolocation mobile.		RADIOLOCATION.
9300-9500	RADIONAVIGATION. (US66) (US71) Meteorological aids. (US67) Radiolocation. (US51)			
9500-10000	Radiolocation.	Radiolocation land. Radiolocation mobile.		RADIOLOCATION.
10000-10500	Amateur. Radiolocation. (NG42)	Amateur. Radiolocation land. Radiolocation mobile.		AMATEUR. RADIOLOCATION.
10. 5-10. 55	RADIOLOCATION.	Radiolocation land. Radiolocation mobile.		RADIOLOCATION.
10. 55-10. 565	FIXED.	Fixed.		DOMESTIC PUBLIC.
10. 565-10. 615	FIXED.	Fixed.		DOMESTIC PUBLIC. Digital termination nodal stations.
10. 615-10. 63	FIXED.	Fixed.		DOMESTIC PUBLIC.

10.68–10.7 (405B)	RADIO ASTRONOMY.			10.68–10.7	G, NG. (US74) (US100)
10.7–10.95	FIXED. MOBILE.			10.7–11.7	NG.
10.95–11.2		10.95–11.2	FIXED. FIXED-SATELLITE (Space-to-earth). MOBILE.		
11.2–11.45	FIXED. MOBILE.				
11.45–11.7	FIXED. FIXED-SATELLITE (Space-to-earth). MOBILE.				
11.7–12.2		11.7–12.2 (405BB) (405BC)	BROADCASTING. BROADCASTING-SATELLITE. FIXED. FIXED-SATELLITE (Space-to-earth). MOBILE except aeronautical mobile.	11.7–12.2	NG. (405BC)
12.2–12.5		12.2–12.5	BROADCASTING. FIXED. MOBILE except aeronautical mobile.	12.2–12.75	NG.
12.5–12.75		12.5–12.75	FIXED. FIXED-SATELLITE (Earth-to-space). MOBILE except aeronautical mobile.		
12.75–13.25	FIXED. MOBILE.			12.75–13.25	NG.

10.63–10.68	FIXED.	Fixed.		DOMESTIC PUBLIC. Digital termination user stations.
10.68–10.7	RADIO ASTRONOMY.			
10.7–10.95	FIXED.	Common carrier fixed.		DOMESTIC PUBLIC. (NG41)
10.95–11.2	FIXED. FIXED-SATELLITE. (NG104)	Common carrier fixed. Space.		DOMESTIC PUBLIC. (NG41) FIXED-SATELLITE.
11.2–11.45	FIXED.	Common carrier fixed.		DOMESTIC PUBLIC. (NG41)
11.45–11.7	FIXED. FIXED-SATELLITE. (NG104)	Common carrier fixed. Space.		DOMESTIC PUBLIC. (NG41) FIXED-SATELLITE.
11.7–12.2	BROADCASTING-SATELLITE. FIXED-SATELLITE. Mobile. (NG105)	Common carrier land. Common carrier mobile (except aeronautical mobile). Space.		
12.2–12.5 (NG8) (NG52)	FIXED.	International control. Operational fixed.		
12.5–12.7 (NG8) (NG52)	FIXED. FIXED-SATELLITE.	Earth. International control. Operational fixed.		
12.7–12.75 (NG 53) (NG 118)	FIXED. FIXED-SATELLITE. MOBILE.	Cable television relay. Earth. Television intercity relay. Television STL.		
12.75–13.20 (NG 53) (NG 118)	FIXED. MOBILE.	Cable television relay. Television intercity relay. Television STL.		
12.95–13.2 (NG118)	FIXED. MOBILE.	Television intercity relay. Television pickup. Television STL.		
13.2–13.25	FIXED. MOBILE.			

§ 2.106 Table of Frequency Allocations—Continued

Worldwide		Region 2		United States	
Band (GHz)	Service	Band (GHz)	Service	Band (GHz)	Allocation
1	2	3	4	5	6
13.25–13.4 (407A)	AERONAUTICAL RADIONAVIGATION. (406)			13.25–13.4	G, NG.
13.4–14.0 (407A)	RADIOLOCATION.			13.4–14.0	G, NG. (US110)
14.0–14.3 (407A)	FIXED-SATELLITE (Earth-to-space). RADIONAVIGATION. (408A)			14.0–14.2	G, NG. (US207)
				14.2–14.3	G, NG. (US207)
14.3–14.4	FIXED-SATELLITE (Earth-to-space). RADIONAVIGATION. SATELLITE. (408A)			14.3–14.4	G, NG. (US206) (US207)
14.4–14.5 (408B) (408C)	FIXED. FIXED-SATELLITE (Earth to-space). MOBILE.			14.4–14.5	G, NG. (US203) (US207)
14.5–15.35 (408B) (408C)	FIXED. MOBILE.			14.5–15.35	G. (US203) (US211)
15.35–15.4 (409C)	RADIO ASTRONOMY.			15.35–15.4	G, NG. (US74) (US100)
15.4–15.7	AERONAUTICAL RADIONAVIGATION. (352A) (352B)			15.4–15.7	G, NG. (352A) (352B) (US118) (US211)
15.7–17.7	RADIOLOCATION.			15.7–17.7	G, NG. (US110)
17.7–19.7	FIXED. FIXED-SATELLITE (Space-to-earth). MOBILE.			17.7–19.7	NG.

Federal Communications Commission				
Band (GHz) 7	Service 8	Class of station 9	Frequency (GHz) 10	Nature { OF SERVICES / of stations 11
13.25–13.4	AERONAUTICAL RADIONAVIGATION (406). Space research.			Airborne doppler radar. Earth.
13.4–14.0	Radiolocation. Space research.	Earth. Radiolocation land. Radiolocation mobile.		
14.0–14.2	FIXED-SATELLITE. RADIONAVIGATION. Space research.	Earth. Radionavigation land. Radionavigation mobile.		
14.2–14.3	FIXED-SATELLITE. RADIONAVIGATION.	Earth. Radionavigation land. Radionavigation mobile.		
14.3–14.4	FIXED-SATELLITE. RADIONAVIGATION SATELLITE.	Earth.		
14.4–14.5	FIXED-SATELLITE. Space research.	Earth. Space.		FIXED-SATELLITE (Earth-to-space). SPACE RESEARCH (Space-to-earth).
15.35–15.4	RADIO ASTRONOMY.			
15.4–15.7	AERONAUTICAL RADIONAVIGATION.			
15.7–17.7	Radiolocation.	Radiolocation land. Radiolocation mobile.		RADIOLOCATION.
17.7–18.36	FIXED. FIXED-SATELLITE. MOBILE.	Fixed. Mobile. Space.		Domestic fixed public. FIXED-SATELLITE.

19.7–21.2 (409E)	FIXED-SATELLITE (Space-to-earth).			19.7–20.2	NG.
				20.2–21.2	G.
21.2–22.0	EARTH EXPLORATION-SATELLITE (Space-to-earth). FIXED. MOBILE.			21.2–22.0	G, NG.
22.0–22.5 (410A)	FIXED. MOBILE.			22.0–23.6	G, NG. (410A)
22.5–23.0		22.5–23.0	FIXED. MOBILE.		
23.0–23.6	FIXED. MOBILE.				
23.6–24.0	RADIO ASTRONOMY.			23.6–24.0	G, NG. (US72) (US74)
24.0–24.05 (410C)	AMATEUR. AMATEUR-SATELLITE.			24.0–24.05	AMATEUR. AMATEUR-SATELLITE. (US72) (US202) (US211)
24.05–24.25 (410C)	RADIOLOCATION. Amateur.			24.05–24.25	G, NG. (US72) (US110) (US202)
24.25–25.25	RADIONAVIGATION. (411)			24.25–25.25	G, NG. (US72)
25.25–27.5	FIXED. MOBILE.			25.25–27.5	G.
27.5–29.5	FIXED. FIXED-SATELLITE (Earth-to-space). MOBILE.			27.5–29.5	NG.
29.5–31.0	FIXED-SATELLITE (Earth-to-space).			29.5–30.0	NG.
				30.0–31.0	G.

18.36-19.04 (NG106)	FIXED. FIXED-SATEL-LITE. MOBILE.	Fixed. Mobile. Space.		Domestic fixed public. Operational fixed. FIXED-SATEL-LITE.
19.04-19.70	FIXED. FIXED-SATEL-LITE. MOBILE.	Fixed. Mobile. Space.		Domestic fixed public. FIXED-SATEL-LITE.
19.7-20.2	FIXED-SATEL-LITE.	Space.		
21.2-22.0 (NG107)	EARTH EXPLO-RATION-SATEL-LITE. FIXED. MOBILE.	Fixed. Mobile except aero-nautical mobile. Space.		Domestic fixed public. Operational fixed. EARTH EXPLO-RATION-SATEL-LITE.
22.0-23.6 (NG107)	FIXED. MOBILE.	Fixed. Mobile except aero-nautical mobile.		Operational fixed. Domestic fixed public.
23.6-24.0	RADIO ASTRONOMY.			
24.0-24.05	AMATEUR. AMATEUR-SATELLITE.	Amateur. Earth. Space.		
24.05-24.25	Amateur. Radiolocation.	Amateur. Radiolocation land. Radiolocation mobile.	24.125	Industrial, scientific and medical equipment.
24.25-25.25	RADIONAVIGA-TION. (411)			
27.5-29.5	FIXED. FIXED-SATEL-LITE. MOBILE.	Common carrier fixed. Common carrier land. Common carrier mobile (except aeronautical mobile). Earth.		
29.5-30.0	FIXED-SATEL-LITE.	Earth.		

§ 2.106 Table of Frequency Allocations—Continued

Worldwide		Region 2		United States	
Band (GHz) 1	Service 2	Band (GHz 3	Service 4	Band (GHz) 5	Allocation 6
31.0–31.3 (412H) (412I)	FIXED. MOBILE. Space research.			31.0–31.2	NG. (US100)
				31.2–31.5	G, NG. (US74) (US100)
31.3–31.5 (412A)	RADIO ASTRONOMY.				
31.5–31.8		31.5–31.8 (405C)	SPACE RESEARCH.	31.5–31.8	G, NG. (US100) (US211)
31.8–32.3 (412B)	RADIONAVIGATION. Space research.			31.8–33.4	G, NG. (US100)
32.3–33.0	RADIONAVIGATION.				
33.0–33.4		33.0–33.4	RADIONAVIGATION. (412F)		
33.4–34.2 (412G)	RADIOLOCATION.			33.4–36.0	G, NG. (412D) (US100) (US110)
34.2–35.2 (412C) (412D)	RADIOLOCATION. Space research.				
35.2–36.0	RADIOLOCATION.				
36.0–40.0 (391A) (412E)	FIXED. MOBILE.			36.0–38.6	G. (US100)
				38.6–40.0	NG.
40.0–41.0	FIXED-SATELLITE (Space-to-earth).			40.0–41.0	G, NG
41.0–43.0	BROADCASTING-SATELLITE.			41.0–43.0	G, NG
43.0–48.0	AERONAUTICAL MOBILE-SATELLITE. AERONAUTICAL RADIONAVIGATION-SATELLITE. MARITIME MOBILE-SATELLITE.			43.0–48.0	G, NG

Federal Communications Commission

Band (GHz)	Service	Class of station	Frequency (GHz)	Nature of SERVICES of stations
7	8	9	10	11
31.0–31.2	FIXED. MOBILE.			
31.2–31.5	RADIO ASTRONOMY.			
31.5–31.8	SPACE RESEARCH.			
31.8–33.4	RADIONAVIGATION. (US69)			
33.4–36.0	Radiolocation.	Radiolocation land. Radiolocation mobile.		RADIOLOCATION.
38.6–40.0	FIXED. MOBILE.			
40.0–41.0	FIXED. FIXED-SATELLITE. MOBILE.	Space.		
41.0–43.0	FIXED. BROADCASTING-SATELLITE. MOBILE.	Space.		
43.0–48.0	AERONAUTICAL MOBILE. AERONAUTICAL MOBILE-SATELLITE. MARITIME-MOBILE. MARITIME-MOBILE SATELLITE.			

	MARITIME RADIONAVIGATION SATELLITE.				
48.0–50.0	Not Allocated.			48.0–50.0	G, NG
50.0–51.0	FIXED-SATELLITE. (Earth-to-space).			50.0–51.0	G, NG
51.0–52.0	EARTH EXPLORATION-SATELLITE. SPACE RESEARCH.			51.0–52.0	G, NG
52.0–54.25 (412J)	SPACE RESEARCH (Passive).			52.0–54.25	G, NG 412J
54.25–58.2	INTER-SATELLITE.			54.25–58.2	G, NG
58.2–59.0 (412J)	SPACE RESEARCH (Passive).			58.2–59.0	G, NG 412J
59.0–64.0	INTER-SATELLITE.			59.0–64.0	G, NG
64–0–65.0 (412J)	SPACE RESEARCH (Passive).			64.0–65.0	G, NG 412J
65.0–66.0	EARTH EXPLORATION-SATELLITE. SPACE RESEARCH.			65.0–66.0	G, NG

	AERONAUTICAL RADIONAVIGATION. AREONAUTICAL RADIONAVIGATION-SATELLITE. MARITIME RADIONAVIGATION. MARITIME RADIONAVIGATION-SATELLITE.			
48.0–50.0	RADIOLOCATION. Amateur.			
50.0–51.0	FIXED. FIXED-SATELLITE. MOBILE.	Earth.		
51.0–52.0	EARTH EXPLORATION-SATELLITE. SPACE RESEARCH.			
52.0–54.25	SPACE RESEARCH (Passive).			
54.25–58.2	FIXED. MOBILE (except aeronautical mobile).			
58.2–59.0	SPACE RESEARCH (Passive).			
59.0–64.0	FIXED. MOBILE (except aeronautical mobile).			
64.0–65.0	SPACE RESEARCH (Passive).			
65.0–66.0	EARTH EXPLORATION-SATELLITE. SPACE RESEARCH.			

§ 2.106 Table of Frequency Allocations—Continued

Worldwide		Region 2		United States	
Band (GHz)	Service	Band (GHz)	Service	Band (GHz)	Allocation
1	2	3	4	5	6
66.0–71.0	AERONAUTICAL MOBILE-SATELLITE. AERONAUTICAL RADIONAVIGATION-SATELLITE. MARITIME MOBILE-SATELLITE. MARITIME RADIONAVIGATION-SATELLITE.			66.0–71.0	G, NG
71.0–84.0	Not Allocated.			71.0–76.0	G, NG
				76.0–84.0	G, NG
84.0–86.0	BROADCASTING-SATELLITE.			84.0–86.0	G, NG (US211)
86.0–92.0 (412J)	RADIO ASTRONOMY. SPACE RESEARCH (Passive).			86.0–92.0	G, NG. (412J) (US74)
92.0–95.0	FIXED-SATELLITE (Earth-to-space).			92.0–93.0	G, NG
				93.0–95.0	G, NG
95.0–101.0	AERONAUTICAL MOBILE-SATELLITE. AERONAUTICAL RADIO-NAVIGATION-SATEL-			95.0–101.0	G, NG

Federal Communicatio				
Band (GHz)	Service	Class of station	Frequency (GHz)	Nature of stations {OF SERVICES
7	8	9	10	11
66.0–71.0	AERONAUTICAL MOBILE. AERONAUTICAL MOBILE-SATELLITE. MARITIME MOBILE. MARITIME MOBILE-SATELLITE. AERONAUTICAL RADIONAVIGATION. AERONAUTICAL RADIONAVIGATION-SATELLITE. MARITIME RADIONAVIGATION. MARITIME RADIONAVIGATION-SATELLITE.			
71.0–76.0	RADIOLOCATION. Amateur.			
76.0–84.0	FIXED. MOBILE.			
84.0–86.0	FIXED. BROADCASTING-SATELLITE. MOBILE.			
86.0–92.0	RADIO ASTRONOMY. SPACE RESEARCH (Passive).			
92.0–93.0	FIXED. MOBILE.			
93.0–95.0	FIXED. FIXED-SATELLITE. MOBILE.	Earth.		
95.0–101.0	AERONAUTICAL MOBILE. AERONAUTICAL MOBILE-SATELLITE. MARITIME MOBILE.			

	LITE. MARITIME MOBILE-SATELLITE. MARITIME RADIO-NAVIGATION-SATELLITE.				
101.0–102.0 (412J)	SPACE RESEARCH (Passive).			101.0–102.0	G, NG (412J)
102.0–105.0	FIXED-SATELLITE (Space-to-earth).			102.0–103.0	G, NG
				103.0–105.0	G, NG
105.0–130.0 (412K)	INTER-SATELLITE.			105.0–110.0	G, NG
				110.0–117.5	G, NG
				117.5–122.5	G, NG
				122.5–130.0	G, NG (US211)
130.0–140.0 (412J)	RADIO ASTRONOMY. SPACE RESEARCH (Passive).			130.0–140.0	G, NG (412J) (US74)
140.0–142.0	FIXED-SATELLITE. (Earth-to-space).			140–0–141.0	G, NG
				141.0–142.0	G, NG

	MARITIME MOBILE-SATELLITE. AERONAUTICAL RADIONAVIGATION. AERONAUTICAL RADIONAVIGATION-SATELLITE. MARITIME RADIONAVIGATION. MARITIME RADIONAVIGATION-SATELLITE.			
101. 0–102. 0	SPACE RESEARCH (Passive).			
102. 3–103. 0	FIXED. MOBILE.			
103. 0–105. 0	FIXED. FIXED-SATELLITE. MOBILE.	Space.		
105. 0–110. 0	FIXED. MOBILE (except Aeronatuical Mobile).			
110. 0–117. 5	FIXED. INTER-SATELLITE. MOBILE (except Aeronautical Mobile).			
117. 5–122. 5	FIXED. MOBILE (except Aeronautical Mobile).			
122. 5–130. 0	FIXED. INTER-SATELLITE. MOBILE (except Aeronautical Mobile).			
130. 0–140. 0	RADIO ASTRONOMY. SPACE RESEARCH (Passive).			
140–0–141. 0	FIXED. MOBILE.			
141. 0–142. 0	FIXED. FIXED-SATELLITE. MOBILE.	Earth.		

§ 2.106 Table of Frequency Allocations—Continued

World Wide		Region 2		United States	
Band (GHz)	Service	Band (GHz)	Service	Band (GHz)	Allocation
1	2	3	4	5	6
142.0–150.0	AERONAUTICAL MOBILE-SATELLITE. AERONAUTICAL RADIO-NAVIGATION-SATELLITE. MARITIME MOBILE-SATELLITE. MARITIME RADIO-NAVIGATION-SATELLITE.			142.0–150.0	G, NG
150.0–152.0	FIXED-SATELLITE. (Space-to-earth).			150.0–151.0	G, NG
				151.0–152.0	G, NG
152.0–170.0	Not Allocated.			152.0–165.0	G, NG
				165.0–170.0	G, NG
170.0–182.0	INTER-SATELLITE.			170.0–175.0	G, NG
				175.0–182.0	G, NG
182.0–185.0 (412J)	SPACE RESEARCH. (Passive).			182.0–185.0	G, NG 412J
185.0–190.0	INTER-SATELLITE.			185.0–189.0	G, NG

Band (GHz) 7	Service 8	Class of Station 9	Frequency (GHz) 10	Nature {OF SERVICES of stations} 11
142.0–150.0	AERONAUTICAL MOBILE. AERONAUTICAL MOBILE-SATELLITE. MARITIME MOBILE. MARITIME MOBILE-SATELLITE. AERONAUTICAL RADIONAVIGATION. AERONAUTICAL RADIONAVIGATION-SATELLITE. MARITIME RADIONAVIGATION. MARITIME RADIONAVIGATION-SATELLITE.			
150.0–151.0	FIXED. MOBILE.			
151.0–152.0	FIXED. FIXED-SATELLITE. MOBILE.	Space.		
152.0–165.0	FIXED. MOBILE.			
165.0–170.0	RADIOLOCATION. Amateur.			
170.0–175.0	FIXED. MOBILE (except Aeronautical Mobile).			
175.0–182.0	FIXED. INTER-SATELLITE. MOBILE (except Aeronautical Mobile).			
182.0–185.0	SPACE RESEARCH (Passive).			
185.0–189.0	FIXED. INTER-SATELLITE. MOBILE (except Aeronautical Mobile).			

				189.0–190.0	G, NG
190.0–200.0	AERONAUTICAL MOBILE-SATELLITE. AERONAUTICAL RADIO NAVIGATION SATELLITE. MARITIME MOBILE-SATELLITE. MARITIME RADIO-NAVIGATION-SATELLITE.			190.0–200.0	G, NG
200.0–220.0	Not Allocated.			200.0–220.0	G, NG
220.0–230.0	FIXED-SATELLITE.			220.0–230.0	G, NG US 211
230.0–240.0 (412J)	RADIO ASTRONOMY. SPACE RESEARCH (Passive).			230.0–240.0	G, NG 412J US74
240.0–250.0	Not Allocated.			240.0–250.0	G, NG
250.0–265.0	AERONAUTICAL MOBILE-SATELLITE. AERONAUTICAL RADIO-NAVIGATION-SATELLITE. MARITIME MOBILE-SATELLITE. MARITIME RADIO-NAVIGATION-SATELLITE.			250.0–265.0	G, NG
265.0–275.0	FIXED-SATELLITE.			265.0–275.0	G, NG
275.0–300.0	Not Allocated.			275.0–300.0	G, NG
Above 300.0	Not Allocated.			Above 300.0	G, NG

189.0–190.0	FIXED. MOBILE (except Aeronautical Mobile).			
190.0–200.0	AERONAUTICAL MOBILE. AERONAUTICAL MOBILE SATELLITE. MARITIME MOBILE. MARITIME-MOBILE-SATELLITE. AERONAUTICAL RADIONAVIGATION. AERONAUTICAL RADIONAVIGATION-SATELLITE. MARITIME RADIONAVIGATION. MARITIME RADIONAVIGATION-SATELLITE.			
200.0–230.0	FIXED. MOBILE.			
220.0–230.0	FIXED. FIXED-SATELLITE. MOBILE.			
230.0–240.0	RADIO ASTRONOMY. SPACE RESEARCH (Passive).			
240.0–250.0	RADIOLOCATION. Amateur.			
250.0–265.0	AERONAUTICAL MOBILE. AERONAUTICAL MOBILE-SATELLITE. MARITIME MOBILE. MARITIME MOBILE-SATELLITE. AERONAUTICAL RADIONAVIGATION.			
265.0–275.0	FIXED. FIXED-SATELLITE. MOBILE.			
275.0–300.0	FIXED. MOBILE.			
Above 300.0	Amateur.			

Index

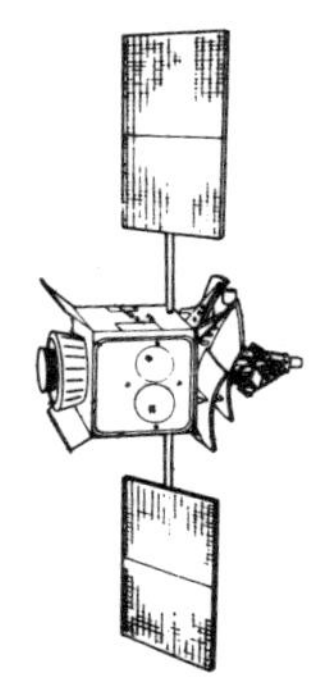

Index